The Receptors

Volume 25

Series editor

Giuseppe di Giovanni

For further volumes:
http://www.springer.com/series/7668

Jeanelle Portelli · Ilse Smolders

Editors

Central Functions of the Ghrelin Receptor

Humana Press

Editors
Jeanelle Portelli
Department of Pharmaceutical Chemistry
Center for Neurosciences, Drug Analysis
  and Drug Information
Vrije Universiteit Brussel
Brussels
Belgium

and

Laboratory for Clinical and Experimental
  Neurophysiology, Neurobiology and
  Neuropsychology
Department of Neurology
Institute for Neuroscience
Ghent University Hospital
Gent
Belgium

Ilse Smolders
Department of Pharmaceutical Chemistry
Center for Neurosciences, Drug Analysis
  and Drug Information
Vrije Universiteit Brussel
Brussels
Belgium

ISBN 978-1-4939-0822-6   ISBN 978-1-4939-0823-3   (eBook)
DOI 10.1007/978-1-4939-0823-3
Springer New York Heidelberg Dordrecht London

Library of Congress Control Number: 2014937692

Humana Press is a brand of Springer
Springer is part of Springer Science+Business Media (www.springer.com)

Foreword

It is my great pleasure and honor to write the Preface of this book dedicated to Ghrelin.

Fifteen years have past since the discovery of Ghrelin. During these years a lot of research has been done to elucidate the physiological functions of Ghrelin, not only a mere growth-hormone-releasing hormone, but also an important appetite regulator, energy conservator, and sympathetic nerve suppressor. At present, Ghrelin is the only circulating orexigenic hormone secreted from a peripheral organ and that acts on the hypothalamic arcuate nucleus, the regulatory region of appetite.

Although the discovery of Ghrelin is dated in 1999, it has a longer history since Dr. Bowers discovered the first growth hormone secretagogue in 1976, paving the way to identify the growth hormone secretagogue receptor, which was the key strategic protein for the discovery of Ghrelin.

I remember very well the day that we began the search for Ghrelin. It was April 7, 1998. That night I found a paper in Science, which reported the identification of the growth hormone secretagogue receptor. I had previously read the manuscript when it was published in the Science journal edition of August 1996. My first impression of the manuscript was why did the prestigious journal Science decide to publish the cloning of a growth hormone releasing peptide receptor. I remember feeling confused with the fact that the cloning of the growth hormone releasing peptide receptor, that is the receptor for another growth hormone releasing peptide from hypothalamus, had been already published. I read the paper about the growth hormone secretagogue receptor very carefully and had confidence that this receptor was a very good target to search for its endogenous ligand.

There have been at least five major breakthroughs in Ghrelin research. The first breakthrough, of course, is the discovery of growth hormone secretagogue by Dr. Bowers. I was deeply impressed to know that my mentor, Dr. Hisayuki Matsuo got acquainted with Dr. Bowers, since Matsuo and Bowers worked in Tulane University with Dr. Andrew Schally, a Nobel prize winner for his discovery of hypothalamic peptide hormones. The second breakthrough was the identification of the growth hormone secretagogue receptor. The work performed by Dr. Howard from Merck research laboratories was not for the faint hearted and required elegant techniques for cloning the receptor. Without the identification of the growth hormone secretagogue receptor, Ghrelin would not have been discovered.

The third breakthrough was the discovery of Ghrelin by my group. The tissue, from which Ghrelin was discovered, was surprisingly in stomach. These results suggest that stomach is not only a digestive organ but also an endocrine organ that secretes growth hormone releasing peptide. The fourth breakthrough was the role of Ghrelin as an orexigenic peptide from peripheral tissue. These results were reported almost at the same time by different independent groups. The fifth breakthrough was the identification of Ghrelin O-acyltransferase (GOAT), Ghrelin O-acyltransferase, which is an acyltransferase specific for acyl-modification of Ghrelin. The identification of GOAT was performed by two independent groups: one by Drs. Brown and Goldstein, the Nobel prize winners from Texas University, and another by Dr. Gutierrez from Eli Lilly. All these breakthroughs progressed the research of Ghrelin and contributed to more than 6,000 published papers on Ghrelin.

Finally, I hope that this book will provide the readers with an up-to-date knowledge on the role of Ghrelin in the central nervous system and attract many researchers to join the study of Ghrelin.

Kurume   Masayasu Kojima

Preface

In 1996, another G-protein coupled receptor (GPCR) was discovered which was added to the ever-increasing list of the seven-transmembrane receptor class: the growth hormone secretagogue receptor. The popularity of this receptor took a U-turn after 1999 when its endogenous ligand Ghrelin was discovered as a result of what is now known as a classical case of reverse pharmacology. The vast pleiotropic physiological properties this ligand presented following its binding to the growth hormone secretagogue receptor led to the latter to be nicknamed 'the Ghrelin receptor'. This is detailed by Prof. Kojima in his foreword, for which we are extremely honored to have as an introduction to our book. The Ghrelin receptor is located in various central and peripheral organs, and is present in different species, which has allowed numerous scientists from entirely different fields to feverishly understand this receptor system. It has taken years for Ghrelin researchers to start understanding the complicated nature of the Ghrelin receptor, having properties that few other GPCRs encompass. This is not a straightforward receptor system, and this was what compelled us to bring together this book that solely focuses on the Ghrelin receptors present in the central nervous system.

The sole aim of this book was to congregate the known different roles of Ghrelin receptors present in the central nervous system, together with a detailed explanation on the intrinsic properties of the receptor itself. The 13 different chapters in this book, each penned by experts in the field, give a complete overview of what is known to date with regards to this receptor in the brain. This concise gathering is aimed as a valuable reference for students, neuroscientists, pharmacologists, and physicians who are working in the Ghrelin receptor field or else are interested in the potential of this receptor axis in the clinical setting.

We would like to thank Springer and its publishing editor for this series for giving us the opportunity to develop and publish this book as part of their 'The Receptors' series. Last but not least, we would also like to express our sincere appreciation to all the chapter authors. This book would not exist were it not for the efforts of all authors who enthusiastically contributed the chapters of this book, for which we are very grateful.

Brussels, Ghent, Belgium   Jeanelle Portelli
Brussels, Belgium   Ilse Smolders

Contents

Contributors

Zane B. Andrews Department of Physiology, Monash University, Clayton, VIC, Australia

Jessica Coppens Department of Pharmaceutical Chemistry, Center for Neurosciences, Drug Analysis and Drug Information, Vrije Universiteit Brussel, Brussels, Belgium

John F. Cryan Food for Health Ireland, University College Cork, Cork, Ireland; Laboratory of Neurogastroenterology, Alimentary Pharmabiotic Centre, University College Cork, Cork, Ireland; Deparment of Anatomy and Neuroscience, Western Gateway Building, University College Cork, Cork, Ireland

Yukari Date Frontier Science Research Center, University of Miyazaki, Miyazaki, Japan

Timothy G. Dinan Food for Health Ireland, University College Cork, Cork, Ireland; Laboratory of Neurogastroenterology, Alimentary Pharmabiotic Centre, University College Cork, Cork, Ireland; Department of Psychiatry, University College Cork, Cork, Ireland

Martin Dresler Max Planck Institute of Psychiatry, Munich, Germany

Jörgen A. Engel Department of Pharmacology, Institute of Neuroscience and Physiology, The Sahlgrenska Academy at the University of Gothenburg, Gothenburg, Sweden

Allison A. Feduccia Section on Clinical Psychoneuroendocrinology and Neuropsychopharmacology, NIAAA and NIDA National Institutes of Health, Bethesda, MD, USA

Birgitte Holst Department of Neuroscience and Pharmacology and NNF Centre for Basic Metabolic Research, University of Copenhagen, Copenhagen N, Denmark

Elisabet Jerlhag Department of Pharmacology, Institute of Neuroscience and Physiology, The Sahlgrenska Academy at the University of Gothenburg, Gothenburg, Sweden

Levente Kapás Washington, Wyoming, Alaska, Montana and Idaho (WWAMI) Medical Education Program and Department of Integrative Physiology and Neuroscience, Sleep and Performance Research Center, Washington State University, Spokane, WA, USA

Nicolas Kunath Max Planck Institute of Psychiatry, Munich, Germany

Alessandro Laviano Department of Clinical Medicine, Sapienza University, Rome, Italy

Lorenzo Leggio Section on Clinical Psychoneuroendocrinology and Neuropsychopharmacology, NIAAA and NIDA National Institutes of Health, Bethesda, MD, USA

Andreas Nygaard Madsen Department of Neuroscience and Pharmacology and NNF Centre for Basic Metabolic Research, University of Copenhagen, Copenhagen N, Denmark

Alessia Mari Department of Clinical Medicine, Sapienza University, Rome, Italy

Ann Massie Department of Pharmaceutical Chemistry, Center for Neurosciences, Drug Analysis and Drug Information, Vrije Universiteit Brussel, Brussels, Belgium

Jacek Mokrosiński Department of Neuroscience and Pharmacology and NNF Centre for Basic Metabolic Research, University of Copenhagen, Copenhagen N, Denmark

Wolfgang H. Oertel Department of Neurology, Philipps-University Marburg, Marburg, Germany

Mario Perello Laboratory of Neurophysiology, Multidisciplinary Institute of Cell Biology, La Plata, Buenos Aires, Argentina

Jeanelle Portelli Department of Pharmaceutical Chemistry, Center for Neurosciences, Drug Analysis and Drug Information, Vrije Universiteit Brussel, Brussels, Belgium; Laboratory for Clinical and Experimental Neurophysiology, Neurobiology and Neuropsychology, Department of Neurology, Institute for Neuroscience, Ghent University Hospital, Gent, Belgium

Jesica Raingo Laboratory of Electrophysiology of the Multidisciplinary Institute of Cell Biology (IMBICE), Argentine Research Council (CONICET) and Scientific Research Commission of the Province of Buenos Aires (CIC-PBA), La Plata, Buenos Aires, Argentina

Harriët Schellekens Food for Health Ireland, University College Cork, Cork, Ireland; School of Pharmacy, University College Cork, Cork, Ireland

Ilse Smolders Department of Pharmaceutical Chemistry, Center for Neurosciences, Drug Analysis and Drug Information, Vrije Universiteit Brussel, Brussels, Belgium

Sarah J. Spencer School of Health Sciences and Health Innovations Research Institute (HIRi), RMIT University, Melbourne, VIC, Australia

Romana Stark Department of Physiology, Monash University, Clayton, VIC, Australia

Éva Szentirmai Washington, Wyoming, Alaska, Montana and Idaho (WWAMI) Medical Education Program and Department of Integrative Physiology and Neuroscience, Sleep and Performance Research Center, Washington State University, Spokane, WA, USA

Marcus M. Unger Department of Neurology, Saarland University, Homburg, Germany

Abbreviations

α−MSH	α-melanocyte-stimulating hormone
5-HT	Serotonin
ACTH	Adrenocorticotropic hormone
AgRP	Agouti-related peptide
AMPK	AMP-activated protein kinase
AP	Area postrema
ARC	Arcuate nucleus
ASP	Agouti-signaling peptide
ATP	Adenosine triphosphate
AVP	Arginine Vasopressin
BBB	Blood–brain barrier
BMI	Body mass index
Ca^{2+}	Calcium
CaM	Calmodulin
CaMKK	Calmodulin-dependent protein kinase kinases
CAMP	Cyclic adenosine monophosphate
CART	Cocaine amphetamine-regulated transcript
CCK	Cholecystokinin
CNS	Central nervous system
CPP	Conditioned place preference
CREB	CAMP response element-binding protein
CRH	Corticotropin-releasing hormone
CSDS	Chronic social defeat stress
CSF	Cerebrospinal fluid
DA	Dopamine
DAD1	Dopamine D1 receptor
DAD2	Dopamine D2 receptor
DAG	Diacyl glycerol
DBS	Deep brain stimulation
Des-acyl Ghrelin	Des-acyl Ghrelin
DIO	Diet-induced obesity
DMH	Dorsomedial hypothalamic nucleus
DMN	Dorsomedial nucleus
DMV	Dorsomotor nucleus of the vagus

EDTA	Ethylenediaminetetraacetic acid
EEG	Electroencephalographic
ERK1/2	Extracellular signal-regulated kinases 1 and 2
EWcp	Centrally projecting Edinger–Westphal nucleus
FAA	Food anticipatory activity
FEO	Food-entrainable oscillator
FRET	Fluorescence energy transfer
GABA	γ-aminobutyric acid
GH	Growth hormone
ghr-/-	Ghrelin knockout
Ghrelin receptor	Growth hormone secretagogue receptor
Ghrelin receptor 1a	Growth hormone secretagogue receptor 1a
Ghrelin receptor 1b	Growth hormone secretagogue receptor 1b
GHRL	Preproghrelin gene
GHRP	Growth hormone releasing peptide
GHSR	Growth hormone secretagogue receptor
GHSR	Growth hormone secretagogue receptor gene
GHSs	Growth hormone secretagogues
GI	Gastrointestinal
GIP	Gastric inhibitory polypeptide
GLP-1	Glucagon-like peptide-1
GOAT	Ghrelin O-acyltransferase
GPCR	G-protein coupled receptor
G_q	G-protein q
H. pylori	*Helicobacter pylori*
HFD	High fat diet
HPA	Hypothalamic-pituitary-adrenal
i.c.v.	Intracerebroventricular
ip	Intraperitoneal
i.v.	Intravenous
IP3	Inositol 1,4,5-trisphosphate
IP_3	Inositol triphosphate
KO	Knockout
LDTg	Laterodorsal tegmental area
LHA	Lateral hypothalamic area
LTP	Long term potentiation
MAO-B	Monoamine oxidase B
MAP	Mitogen-activated protein
MC1R	Melanocortin receptor 1
MC3R	Melanocortin 3 receptors
MC4R	Melanocortin 4 receptors
MeA	Medial nucleus of the amygdala
mPFC	Medial prefrontal cortex
MPP+	1-methyl-4-phenylpyridinium
MPTP	Methyl-4-phenyl-1,2,3,6-tetrahydropyridine

mTORC1	Mammalian target of rapamycin 1
Na^+	Sodium
NAc	Nucleus Accumbens
NAD	Noradrenaline
NMDA	N-methyl-D-aspartate
NMUR1	Neuromedin receptor 1
NMUR2	Neuromedin receptor 2
NO	Nitric Oxide
NOS	Nitric Oxide Syntase
NPY	Neuropeptide Y
NREMS	Non-rapid eye movement sleep
NTS	Nucleus tractus solitarius
NTSR1	Neurotensin receptor 1
NTSR2	Neurotensin receptor 2
ORX	Orexin
OXM	Oxyntomodulin
PBN	Parabrachial nucleus
PC	Prohormone convertase
PD	Parkinson's Disease
PKA	Protein kinase A
PKC	Protein kinase C
POMC	Pro-opiomelanocortin
PVN	Paraventricular nucleus of the hypothalamus
PWS	Prader-Willi Syndrome
PYY	Peptide tyrosine-tyrosine
PYY	Peptide YY
REM	Rapid eye movement
REMS	Rapid eye movement sleep
SNPs	Single nucleotide polymorphisms
sc	Subcutaneous
SP	Substance P
SRE	Serum response element
STN	Subthalamic Nucleus
SWA	Slow-wave activity of the EEG
TM	Transmembrane
$TP\alpha$	Thromboxane A_2
TSH	Thyroid-stimulating hormone
UCP2	Uncoupling protein 2
VMH	Ventromedial hypothalamic nucleus
VTA	Ventral tegmental area
VMN	Ventromedial nucleus
WHO	World Health Organization
WT	Wildtype

Part I
The Ghrelin Receptor Isoforms

Constitutive Activity of the Ghrelin Receptor

Jacek Mokrosiński, Andreas Nygaard Madsen and Birgitte Holst

Abstract Cloning and characterization of the ghrelin receptor as a 7-transmembrane (7TM), G-protein-coupled receptor (GPCR) was first reported by Howard and his co-workers (1996). The ghrelin receptor was initially described as a growth hormone secretagogue receptor since (GHSR) this was the most well-established physiological function at that time. The natural endogenous agonist remained unknown until Kojima and his co-workers discovered (1999) the peptide hormone ghrelin. Afterward, the activity of ghrelin receptors was linked primarily with the regulation of appetite, adiposity, and energy expenditure as well as inducing of growth hormone secretion (Davenport et al. 2005; Kojima et al. 2001). Another important milestone in the pharmacological characterization of the ghrelin receptor was the discovery of its constitutive activity (Holst et al. 2003, 2004). This chapter will focus on the molecular basis of this phenomenon and its relevance in health and disease.

Keywords Ghrelin receptor · Constitutive activity · Activation mechanism · Ghrelin · Inverse agonist · Substance P analog

J. Mokrosiński · A. N. Madsen · B. Holst (✉)
Department of Neuroscience and Pharmacology and NNF Centre
for Basic Metabolic Research, University of Copenhagen, Blegdamsvej 3b,
build. 18.5, 2200 Copenhagen N, Denmark
e-mail: holst@sund.ku.dk

J. Mokrosiński
e-mail: jacek@sund.ku.dk

A. N. Madsen
e-mail: anmadsen@sund.ku.dk

J. Portelli and I. Smolders (eds.), *Central Functions of the Ghrelin Receptor*,
The Receptors 25, DOI: 10.1007/978-1-4939-0823-3_1,
© Springer Science+Business Media New York 2014

Ghrelin Receptor

The ghrelin receptor belongs to the rhodopsin-like family of 7TM receptors, also known as the class A receptor family. Like all other members of the superfamily, the ghrelin receptor has seven membrane-spanning α-helical domains linked by three extracellular and three intracellular loops. The N- and C-terminal parts are located extra- and intracellularly, respectively. The binding pocket for endogenous and synthetic agonists and antagonists is located at its extracellular site both within the helical bundle and the extracellular loops, while the signaling effectors interact with the receptor at its intracellular surface. The ghrelin receptor constitutes its own receptor subfamily comprised of receptors for motilin (GPR38), neuromedin (NMUR1 and NMUR2), and neurotensin (NTSR1 and NTSR2) and an orphan receptor GPR39 (Holst et al. 2004).

All members of this subfamily share a similar gene structure and a relatively high level of sequence homology (Holst et al. 2004, 2007b). The crystal structure of agonist-bound NTSR1, a member of the ghrelin receptor subfamily, was resolved recently. It brings insight into understanding this family of peptide receptors and their ligand-binding properties. Since NTSR1 was crystallized in the presence of neurotensin-derived peptide and resembles active conformations of previously shown rhodopsin and β2-adrenergic receptors, this structure might also help to describe the conformational features necessary for the constitutive activation of the ghrelin receptor (White et al. 2012).

The gene encoding the ghrelin receptor is located on chromosome 3, locus 3q26.31 and contains two exons (NCBI Gene ID: 2693). This allows for two alternative splicing variants of the ghrelin receptor; a long variant transcribed into a full-length 7TM receptor (ghrelin receptor 1a) and a short, 3′-truncated variant which encodes a 5TM receptor-like protein (ghrelin receptor 1b). Only the long form of the receptor is functional, whereas the short variant does not exhibit any binding or functional activity in response to ghrelin (Howard et al. 1996). The truncated form of the receptor acts as a dominant-negative mutant by impairing the cell surface expression of the receptor (Leung et al. 2007). It has been shown that the 5TM truncated splice variant of the ghrelin receptor can form heterodimers with the full-length receptor. Dimerization of the non-signaling short variant with the full-length ghrelin receptor in a heterodimer prevents changes in the receptor conformation underlying the activation process. Thus, the long form of receptor cannot activate its downstream signaling pathways. Homodimers composed of two full-length ghrelin receptor protomers are fully functional like monomeric receptors (Mary et al. 2013).

It has been reported that the ghrelin receptor can also exist in heterodimeric complexes with other 7TM receptors, e.g. dopamine D1 and D2, melanocortin MC3, serotonin 5-HT$_{2C}$ receptors as well as selected prostanoid receptor family members (Chow et al. 2008; Jiang et al. 2006; Rediger et al. 2011; Schellekens et al. 2013). The presence of the ghrelin receptor in a heterodimer might modulate the signaling properties of either one of the receptors in the complex or both of them. For example, dimerization of the ghrelin receptor with the dopamine D1 or melanocortin MC3 receptors results in the amplification of the dopamine or melanocortin receptor-mediated cAMP production. At the same time, the dimerization can diminish the ghrelin receptor-specific, ligand-induced, and constitutive signal transduction through the $G\alpha_{q/11}$-protein pathway (Chow et al. 2008; Jiang et al. 2006; Lau et al. 2009; Rediger et al. 2011; Schellekens et al. 2013). This phenomenon might contribute to the broad spectrum of physiological functions mediated by the ghrelin receptor, for example, the neurological control of appetite, the rewarding mechanism and memory performance (Abizaid et al. 2006; Diano et al. 2006; Jerlhag et al. 2009; Perello et al. 2010; Rediger et al. 2012).

In summary, the ghrelin receptor belongs to the rhodopsin-like receptor family and constitutes together with motilin (GPR38), neuromedin (NMUR1 and NMUR2) and neurotensin (NTSR1 and NTSR2), and an orphan receptor GPR39, a small subgroup of the receptors which share some structural and functional features. An alternative splicing of the ghrelin receptor may result in the expression of a nonfunctional 5TM truncated variant of the ghrelin receptor. The ghrelin receptor was shown to form heterodimers with several other 7TM receptors. Heterodimerization may have an impact on the receptor signaling properties, including its constitutive activity.

Receptor Activation Mechanism and Constitutive Signaling

A general model of 7TM receptors signaling, known as the ternary complex model, includes three basic components: ligand, receptor, and G-protein (Fig. 1a). In this model, the ligand is an agonist, such as a peptide hormone, a neurotransmitter, a nucleotide, a fatty acid, or other substance, which binds to the receptor and induces G-protein interaction at the intracellular surface of the receptor. The model assumes that the receptor can adopt distinct conformations defined as inactive (R) and active (R*) G-protein-coupled state (De et al. 1980). The transition between the two activation states requires the receptor to surmount an energy barrier (Gether et al. 1997). Generally, agonist binding is required to overcome this energy barrier; however, some receptors may accommodate the active conformation—the R* state—without any need for an agonist. This has been included in the extended version of the ternary complex model (Fig. 1b) (Lefkowitz et al. 1993; Samama et al. 1993). This phenomenon is known as constitutive receptor activity, and is reflected by an increased basal signaling proportional to increasing receptor expression at the cell surface observed in the absence of the receptor agonist. Accordingly, the occurrence of constitutive signaling can be explained by a smaller activation energy barrier (from R to R*) which can be overcome without the presence of its agonist; however, the agonist can stabilize a conformation of the receptor with an even higher activity level (Deupi and Kobilka 2010; Gether et al. 1997).

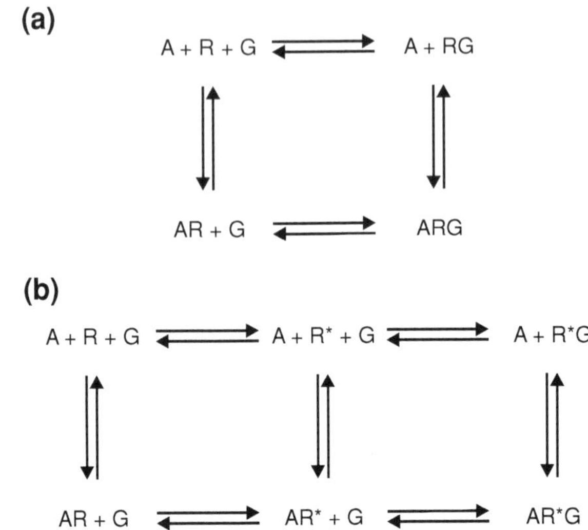

Fig. 1 Schematic representation of the ternary complex model (**a**) and the extended ternary complex model (**b**), where A represents an agonist; R, a receptor in its inactive state; R^*, a receptor in the active state; G, a G-protein [adapted from (Lefkowitz et al. 1993; Samama et al. 1993)]

The constitutive activity is observed only in a few wild-type receptors, including the ghrelin and the MC4 receptors. Increased basal signaling can be induced by naturally occurring mutations and by mutations engineered on the basis of structural interest (Seifert and Wenzel-Seifert 2002). Mutations causing constitutive activity can be obtained either by stabilizing the receptor in its active conformation or by destabilizing structural constraints responsible for low receptor energy that decrease the energy barrier between the R and R* states. Interestingly, amino acid substitutions responsible for the constitutive activity can be found or introduced in almost any region of the receptor. A domain particularly prone to the occurrence of constitutively active mutation is the third intracellular loop. This loop is a part of the receptor—G-protein interface, and therefore mutations in this region might result in conformational changes facilitating the G-protein binding to the receptor (Kjelsberg et al. 1992; Kudo et al. 1996; Ren et al. 1993). Constitutive activity can also be achieved by substitutions within the helical bundle, for example, in TM-VI or -VII, where the mutation is expected to facilitate the active conformation (Steen et al. 2013; Yanagawa et al. 2013). Other mutations resulting in increased basal signaling can be found at the extracellular site of the receptor, suggesting that they mimic the conformational changes induced by ligand binding (Levin et al. 2002; Okada et al. 2004).

The knowledge about the structural basis for activation, e.g., the conformational changes that characterize R versus R*, has greatly increased over the last few years with the help of the crystal structures of the 7TM receptors both in the inactive and active conformation. It has long been known that the rhodopsin-like receptors share characteristic structural features in the transmembrane helical bundle, including a number of conserved sequence motifs, such as DRY in TM-III, CWxP motif in TM-VI, NPxxY (where x can be any amino acid) in TM-VII, and others. The crystal structures reveal that these conserved motifs play a role as micro-switches that facilitate the receptor activation process. Substituting residues in these receptor regions has been described as inducing constitutive activity, and in other cases the substitution of these key residues may abolish the receptor activation. In the $\beta2$ adrenergic receptor, the crystal structure shows that ArgIII:26 from the DRY motif interacts with the neighboring acidic residue in position III:25 in the inactive state, and shifts to interactions with TyrV:24 and the G-protein in the active conformation (Nygaard et al. 2009; Scheerer et al. 2008). Substitutions introduced in the DRY sequence often result in changes in the receptor constitutive signaling due to more favorable stabilization, either of the active or inactive conformation (Case et al. 2008; Jensen et al. 2012; Rovati et al. 2007; Schneider et al. 2010). More generally, changes in interaction patterns within the micro-switches may constitute an important part of the whole receptor conformational shift occurring upon its activation. For example, in the inactive receptor state, the AsnVII:16 residue of the NPxxY motif is oriented toward TM-VI. Various substitutions of the IleVI:05 residue in the histamine H1 receptor located between the DRY motif in TM-III and the NPxxY motif in TM-VII result in constitutive activity. Mutation of Ile in TM-VI into a charged residue (Arg/Lys/Glu) facilitates hydrogen bond interaction, either directly or mediated through free water molecules between AsnVII:16, other polar residues in DRY motif and highly conserved AspII:10. The change of AsnVII:16 orientation from TM-VI toward TM-II and TM-III causes the receptor to adopt its active conformation (Bakker et al. 2008).

The "Global Toggle Switch Model" has been proposed by Schwartz and co-workers (2006) to describe the activation process for 7TM receptors. This comprehensive model assumes that agonist binding to the orthosteric ligand-binding site at the extracellular site of the receptor stabilizes this receptor in its active state where the extracellular ends of TM-VI and TM-VII are moved closer to TM-III. Receptor interaction with a trimetric G-protein is facilitated by an outward movement of the intracellular fragments of the transmembrane helixes, primarily TM-VI and -VII (Elling et al. 2006; Schwartz et al. 2006). A comparison between the inactive and active structures of 7TM receptors resolved using X-ray crystallography shows substantial outward movements of TM-VI and TM-V relative to TM-III (Rasmussen et al. 2011; Scheerer et al. 2008). Rearrangements accompanying the activation process are also observed between TM-III and TM-VII. Helices movements at the extracellular site of the receptor are more subtle compared to those described for the intracellular site. Spatial rearrangements induced by agonist binding mostly concern the same TM domains as the ones shown to facilitate interaction with signal transducers (Katritch et al. 2013). While changes at the receptor—G-protein interface during activation can be considered as a general mechanism, interactions between the receptor and its specific ligand vary substantially due to the multitude of structures of both parties. Thus, spatial rearrangements of the receptor domains induced by ligand binding can be well resolved by crystallography for a particular receptor-ligand pair and might be translatable for the homologous receptors. More profound knowledge concerning conformational rearrangements within transmembrane helices is still lacking for the complete understanding of the receptor activation mechanism.

7TM receptors induce signal transduction through heterotrimeric G-proteins and β-arrestin, which then may activate a broad range of intracellular effectors (Lefkowitz and Shenoy 2005). The ghrelin receptor is primarily coupled to the $G\alpha_{q/11}$ protein, which activates phospholipase C (PLC) and leads to the generation of two secondary messengers: diacyl glycerol (DAG) and inositol 1, 4, 5-tris-phosphate (IP3) (Holst et al. 2003). IP3 released into the cytoplasm may further induce Ca^{2+} signaling pathways. $G\alpha_q$-downstream signaling kinases, such as Ca^{2+}/calmoduline-dependent kinase IV and protein kinase C (PKC), may phosphorylate the cAMP response element-binding protein (CREB) (Matthews et al. 1994; Singh et al. 2001). However, pharmacological profiling of the ghrelin receptor agonists suggests that IP3 turnover and Ca^{2+} mobilization do not represent the same receptor coupling. It is can be speculated that Ca^{2+} mobilization is mediated downstream of the $G\alpha_i$ and/or $G\beta\gamma$ subunit. The ghrelin receptor also couples to the $G\alpha_{12/13}$ protein. This can induce an Rho GTPase signaling pathway resulting in further activation of the serum response element (SRE) (Fig. 2) (Holst et al. 2004; Sivertsen et al. 2011; Holst et al. unpublished observations). β-arrestin mobilization was initially considered as a 7TM receptor signaling suppressing mechanism. However, β-arrestin binding occurs with the receptor present in its active conformation and might result in the activation of other intracellular signaling pathways, for example, the mitogen-activated protein (MAP) kinases cascade. The ghrelin receptor was found to mobilize β-arrestin 2, and to induce the ERK1/2 MAP kinase pathway (Fig. 2) (Holliday et al. 2007; Holst et al. 2004).

In the activation process, a 7TM receptor changes its conformation from an inactive to the active state. Several structural components of the receptor, known as micro-switches, are responsible for stabilization of the active receptor conformation. The transition between the inactive and active conformations requires overcoming of the energy barrier. The active conformation can be adopted by a receptor in the presence of an agonist. Some receptors can also adopt the active conformation without any ligand and this phenomenon is referred to as the constitutive activity.

Structural Features Responsible for the Constitutive Activity of the Ghrelin Receptor

The ghrelin receptor and two other members of its subfamily—NTSR2 and GPR39—exhibit constitutive activity. All these receptors share a structural feature, an aromatic cluster inside the helical bundle, which is comprised of residues

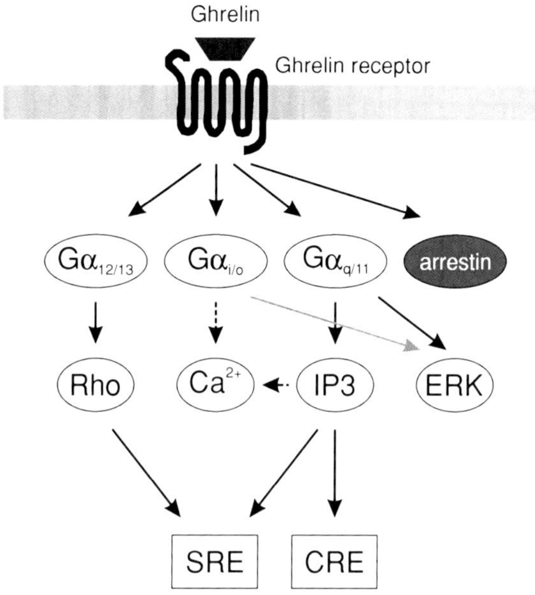

Fig. 2 Schematic overview of the ghrelin receptor signal transduction pathways. The ghrelin receptor couples to $G\alpha_{q/11}$, $G\alpha_{12/13}$, and $G\alpha_{i/o}$ subunit activating their specific downstream signaling pathways; i.e., the inositol phosphates cascade (*IP3*), the intracellular Ca^{2+} mobilization (*Ca^{2+}*), and the RhoA kinase cascade (*Rho*), respectively. Activation of the ghrelin receptor induces also signal transduction through the extracellular signal-regulated kinases (*ERK*) pathway. A physiological response to ghrelin receptor activity is mediated by expression of genes regulated by the serum (*SRE*) and cAMP (*CRE*) response elements. Additionally, a stimulation of the ghrelin receptor with its endogenous agonists ghrelin leads to the β-arrestin mobilization [adapted from (Sivertsen et al. 2013)]

located in positions VI:16, VII:06, and VII:09 (Fig. 3a). It has been suggested that direct interaction between these aromatic side chains stabilizes the receptor in the active conformation by bringing TM-VI and TM-VII closer to each other and pulling them toward TM-III. Mutational studies show that PheVI:16 constitutes a critical role for this aromatic cluster and for the constitutive signaling. Both NTSR2 and the ghrelin receptor carry an aromatic residue in this position, a Tyr and Phe, respectively. However, in both cases the Ala-substitution of this aromatic side chain selectively impairs the constitutive activity. In contrast, GPR39, which exhibits a relatively lower level of constitutive signaling when compared with the ghrelin receptor, has an Asn residue in position VI:16. Substitution of the neutral Asn side chain with an aromatic Phe moiety results in a substantial increase in basal receptor activity, which reaches the same level as that seen for the ghrelin receptor and the NTSR2 (Holst et al. 2004). In a similar manner, the other aromatic residues in TM-VI and TM-VII have been shown by mutational analysis to affect the constitutive signaling (Holst et al. 2004). Additional polar interaction, which may help to bring extracellular ends of TM-III and -VI toward each other and

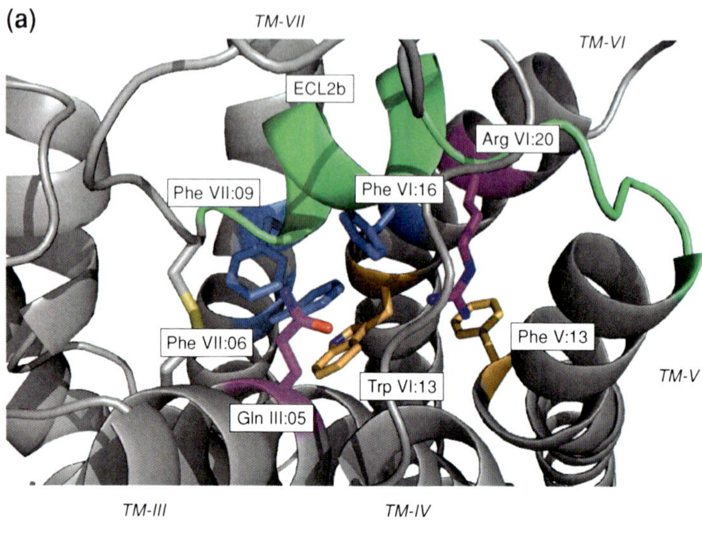

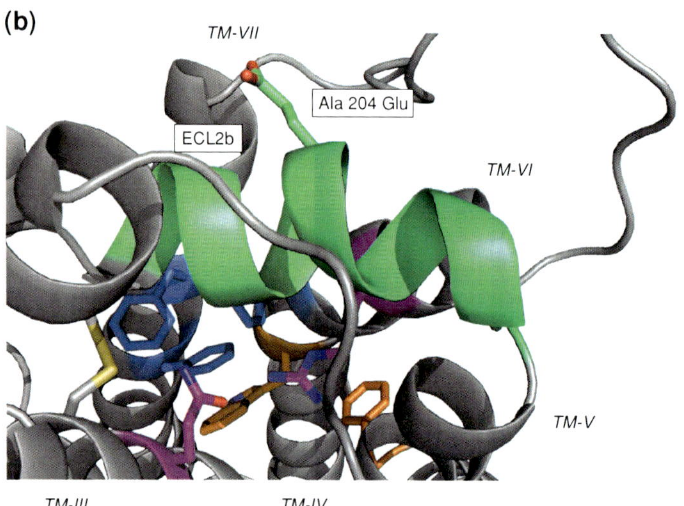

Fig. 3 A fragment of the ghrelin wild-type receptor (**a**) and Ala204Glu mutant (**b**) homology models seen from the extracellular site. Structural features responsible for the constitutive activity were color coded; the PheVI:16, PheVII:06, and VIII:09 residues which form the aromatic cluster are shown in *blue*, the PheV:13 and TrpVI:13 residues which constitute a micro-switch—in *orange*, the GlnIII:05 and ArgVI:20 which stabilize the active conformation through a polar interaction—in magenta, respectively. The extracellular loop 2 (shown in *green*) adopts a flexible conformation in the wild-type receptor (*panel a*), while a naturally occurring Ala204Glu mutation (*panel b*, mutated residue shown as sticks) stabilizes more rigid α-helical structure of the entire extracellular loop fragment from the disulfide bridge with TM-III to the end of TM-V

therefore stabilize the active conformation of ghrelin receptor, is formed by the above-mentioned aromatic cluster together with GlnIII:05 and ArgVI:20 (Holst et al. 2004).

Another structural feature of the ghrelin receptor facilitating its ligand independent signaling is an aromatic interaction between TrpVI:13 and PheV:13 (Fig. 3a) (Holst et al. 2010; Nygaard et al. 2009). Both residues are highly conserved among rhodopsin-like 7TM receptors. TrpVI:13 is part of the CWxP motif and is considered one of the molecular micro-switches which facilitate receptor activation. The indol side chain in position VI:13 is expected to change its conformation as a result of the transition from inactive to active state of the receptor and to stabilize the active conformation by the formation of aromatic interactions with the benzyl ring of PheV:13. Ala-substitutions of both aromatic residues caused loss of the constitutive signaling by the ghrelin receptor while their expression at the cell surface and ligand-binding ability were not significantly affected (Holst et al. 2010).

Several single-point mutations which diminish or eliminate the ghrelin receptor constitutive activity but which do not impair ghrelin-induced signaling have been identified as potential human disease mutations (Liu et al. 2007; Pantel et al. 2006). These residues, reported as being substantial for the agonist-independent receptor basal activity, might indicate structural features that facilitate the transition between the inactive and active states of the receptor. One of such mutations is the Leu-substitution of PheVI:16 (F297L)—the key residue of the previously described aromatic cluster (Holst et al. 2004; Liu et al. 2007). Other variants impairing the constitutive activity were found at the intracellular end of TM-IV— ValIV:02Met (V160 M) and within the extracellular loop 2—Ala204Glu (Liu et al. 2007; Pantel et al. 2006). The molecular mechanism underlying the loss of constitutive activity caused by ValIV:02Met remains to be clarified; however, in the case of Ala204Glu mutation, it has recently been resolved. This particular mutation has previously been shown to be associated with short statue syndrome and potentially also with postpuberty obesity (Pantel et al. 2006). The mutation was found within the extracellular loop 2, the most variable region among all rhodopsin-like receptors. In vitro studies have shown that exchanging the short aliphatic methyl side chain of Ala in position 204 with any charged residue— either positively or negatively—resulted in a loss of constitutive activity as seen for the naturally occurring substitution of Glu. In silico predictions suggest that changes in the secondary structure can explain the surprisingly similar functional response induced both by a negative and positive charge introduced in the extracellular loop 2. In both cases, the significant differences in propensity for an α-helical structure formation between the wild type and the mutated variants of the ghrelin receptor were observed. The charged residue in the middle of the extracellular loop 2 chain (Fig. 3b) induced a rigid α-helix that constrains the distance between TM-III and TM-V (Mokrosinski et al. 2012).

Summarized, comparative analysis of the ghrelin receptor, NTSR2 and GPR39 reveals several structural features important for the relatively high constitutive signaling within this subfamily. The active conformation of the ghrelin receptor was shown to be stabilized in the absence of its agonist by interactions within the aromatic cluster located in TM-VI and TM-VII, the polar interaction between GlnIII:05 and ArgVI:20 and another aromatic interaction between PheV:13 and TrpVI:13. Additionally, the constitutive signaling of the ghrelin receptor is possible because of the flexible structure of the extracellular loop 2. Constrain of the α-helical structure of this domain by the Ala204Glu substitution impairs the ghrelin receptor constitutive activity probably by restricting spontaneous movements of TM-III and TM-V which are spanned by the extracellular loop 2.

Ligands Modulating Constitutive Activity

A term "inverse agonist" describes a ligand that can decrease the constitutive receptor signaling by stabilizing the receptor in its inactive conformation (Costa and Cotecchia 2005). The only endogenous inverse agonists described until now are the agouti-related peptide (AgRP) and the agouti-signaling peptide (ASP) which can inhibit the basal signaling of the melanocortin receptors 1 and 4 (MC1R, MC4R), respectively (Adan 2006). The first reported inverse agonist for the ghrelin receptor was a highly modified substance P analog denoted [D-Arg1, D-Phe5, D-Trp7, 9, Leu11]-substance P (further denoted in this chapter as SP analog), previously shown to be an antagonist for ghrelin receptor (Fig. 4) (Hansen et al. 1999; Holst et al. 2003). By truncation of the peptide sequence, it was found that a heptapeptide D-Phe-Gln-D-Trp-Phe-D-Trp-Leu-Leu (fQwFwLL) was responsible for the inverse agonism of the SP analog. This peptide can be further truncated from the N-terminal site by two amino acids; however, the resulting pentapeptide (wFwLL) at low concentrations acts as a partial agonist and only at higher concentrations retains its inverse agonistic properties (Holst et al. 2006). The addition of positively charged (Arg or Lys), negatively charged (Asp), or aliphatic (Ala) residues at the N-terminus of the pentapeptide turns this ligand into a pure inverse agonist, neutral agonist, or pure partial agonist, respectively (Holst et al. 2006, 2007a).

According to current views on the receptor activation mechanism, agonist binding stabilizes the active conformation of the receptor while an inverse agonist acts in the opposite way, keeping the receptor in its inactive form. The inverse agonist, the SP analog, binds intracellularly in the receptor binding crevice and involves interactions with a broad range of residues located in TM-II, -III, -IV, -V, -VI, and -VII. Moreover, space-generating Ala-substitutions of several residues in TM-IV and -V improve the inverse agonistic properties of the SP analog (Holst et al. 2006). The site of the binding pocket seems to be crucial for the substance P analog derivatives' mode of action. Removal of the polar side chain in position SerIV:16 by Ala-substitution results in the increased efficacy of one of the inverse agonists; K-wFwLL. Agonism of the Ala-extended substance P active core pentapeptide, A-wFwLL, can be modulated by mutation in TM-V, i.e., Ala-substitution of MetV:05 increases its efficacy while ValV:08 Ala and PheV:12 Ala

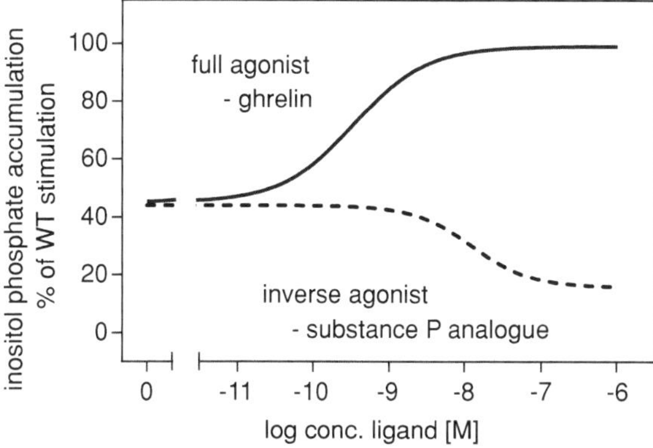

Fig. 4 Dose-response curves of the ghrelin receptor full agonist—ghrelin and inverse agonist—substance P analog. The level of receptor activation was monitored using the inositol phosphate accumulation assay. The ghrelin wild-type receptor was transiently expressed in heterologous mammalian expression system. The constitutive activity level is seen as an increase in the background signaling observed in the absence of an agonist or an inverse agonist as seen for the "0" point on the X axis

mutations eliminate the agonistic properties of this peptide (Holst et al. 2007a; Mokrosinski and Holst 2010). The mode of action of two substance P-derived hexapeptides, K-wFwLL and A-wFwLL, can also be modified by substitutions in TM-III. The inverse agonist, K-wFwLL acts as an agonist in SerIII:08 Ala mutant. Exchange of the Phe III:04 residue (located one helical turn above Ser III:08) with a polar Ser side chain turns the agonist, A-wFwLL into an inverse agonist. A similar swap of the inverse agonism of K-wFwLL into agonism is induced by Ile IV:20 Ala mutation. At the same time, this mutation improves the efficacy of the A-wFwLL peptide from partial into nearly complete agonism (Holst et al. 2007a).

The constitutive activity can be decreased by an inverse agonist. The SP analog, previously shown to be a ghrelin antagonist, is an efficacious inverse agonist of the ghrelin receptor. Truncations and modifications of the SP analog led to discovery of the inverse agonist core peptide and showed that mutations in TM-III, TM-IV, and TM-V can modulate the efficacy or even result in a swap between agonism and inverse agonism of selected SP analog derivatives.

Importance of Constitutive Activity in Vivo

Constitutive activity levels have been described for several 7TM G-protein-coupled receptors either after the introduction of mutations or in the wild-type receptor. Many studies have shown the importance of the constitutive activity in vitro; however, the in vivo importance has mainly been suggested by disease mutations in humans and by a few rodent studies (Arvanitakis et al. 1998; Smit et al. 2007).

One of the most well-described 7TM receptors where a mutation, that induces a high level of constitutive activity, is responsible for the development of a human disease, is the visual receptor for light, where photon absorption causes a conformational change in the light-sensitive rhodopsin molecule, resulting in a phototransduction cascade. It has been demonstrated that mutations in the rhodopsin receptor leading to constitutive activity are associated with rod cell death due to the initiation of apoptosis. This results in night blindness in humans, which is referred to as a rare form of retinitis pigmentosa (Berson 1993; Daiger et al. 2013; Pan et al. 2012; Rivera-De la Parra et al. 2013; Vishnivetskiy et al. 2013).

Another receptor where mutations initiating constitutive activity have been demonstrated to induce a human disease is the glycoprotein hormone receptor for thyroid-stimulating hormone (TSH). Glycoprotein receptors have large extracellular domains, where mutations have been reported to induce constitutive activity. Most commonly, however, the constitutive active mutations are detected in the intracellular loops and in TM-VI. Such mutations exert—even in heterozygote form—a continuous stimulation of growth and differentiation of the thyroid cells, leading to diffuse hyperplasia and hypersecretion of thyroid hormone (Corvilain et al. 2001; Jaeschke et al. 2006).

The wild type MC4 receptor is the best described constitutive active receptor, mainly due to the fact that an endogenous antagonist and inverse agonist AgRP exist. The in vivo importance of the constitutive activity has been substantiated by an elegant study where the AgRP was shown to decrease chronic feeding behavior in mice with a neural-specific knockout of the pro-opiomelanocortin (POMC), eliminating the endogenous agonist melanocyte-stimulating hormone α (α-MSH). This indicates that the inverse agonist properties of AgRP and, accordingly, the constitutive activity of the MC4R is sufficient to modulate feeding behavior (Tolle and Low 2008).

Role of Constitutive Activity of the Ghrelin Receptor in Vivo

The orexigenic hormone ghrelin is known to activate the ghrelin receptor which is highly expressed in the appetite center of the hypothalamus and in the pituitary. Physiologically, ghrelin receptor signaling stimulates growth hormone (GH) secretion as well as body weight regulation in vivo (Holst et al. 2003). In humans, the ghrelin level rises just before a meal and falls again after the meal. Interestingly, the constitutively active ghrelin receptor expression is highly upregulated during fasting at the same time as the ligand ghrelin is upregulated (Kim et al. 2003; Nogueiras et al. 2004; Petersen et al. 2009). This opposite of what is observed to receptors in general, where high concentration of the hormones, is associated with decreased receptor expression and it indicates an important physiological role of the high constitutive activity of the ghrelin receptor independent of its ligand. To study the importance of the constitutive activity of the ghrelin receptor pharmacologically, approaches have been used to selectively lower the constitutive activity. SP analog, the previously mentioned selective inverse agonist of the ghrelin receptor, was given centrally to rats in concentrations where no antagonistic properties were shown and was found to lower food intake and prevent body weight gain (Petersen et al. 2009).

Mutations in the ghrelin receptor locus were analyzed in relation to the increased prevalence of obesity and overweight among white Danes in a general population-based study (Gjesing et al. 2010). Gjesing and her co-workers did not find any association with measures of obesity and overweight in common variations in the ghrelin receptor. Interestingly, they found a rare promoter variant that showed partial co-segregation with obesity and overweight in two pedigrees of whites (Gjesing et al. 2010). This rare mutation was located in the ghrelin receptor promoter region and resulted in an increase in the transcriptional activity of the ghrelin receptor. Consequently, due to the high constitutive activity of the ghrelin receptor, the increased amount of receptors led to an increased signaling independent of the ghrelin. This increased ghrelin receptor signaling could result in a decrease in energy expenditure and appetite regulation in the Danish pedigree (Gjesing et al. 2010).

In another study, the previously mentioned ghrelin receptor mutation Ala204-Glu has been associated with short statue syndrome within two independent families (Pantel et al. 2006). The mutation was located within the extracellular loop 2, and resulted in a lower constitutive activity of the ghrelin receptor without changing ghrelin's capacity to signal through the receptor. Interestingly, there were several postpuberty obese family members indicating an effect of the mutation Ala204Glu on energy homeostasis (Pantel et al. 2006).

Summary and Perspectives

The ghrelin receptor is one of a few 7TM receptors which exhibit a relatively high level of constitutive activity. Based on our current knowledge about receptor structure and conformational dynamics during activation process, the constitutive activity is explained as spontaneous transition from the inactive to the active state possible due to the low energy barrier between these states for this particular receptor. This energy barrier, and therefore the constitutive activity, can be modulated by mutations and receptor-specific ligands. The naturally occurring mutation affecting the constitutive receptor signaling may account for pathogenesis of various diseases. In contrast, development of a potent and efficacious inverse agonist may lead to a new therapeutic strategy. Since the first report on the ghrelin receptor constitutively activity published in 2003 by Holst and her co-workers, we learned a lot concerning the molecular mechanisms underlying occurrence of this phenomenon and its physiological importance. Even though, the ghrelin receptor and its constitutive activity in particular, remain a fascinating subject of both basic pharmacological studies and applied drug discovery programs.

References

Abizaid A, Liu ZW, Andrews ZB et al (2006) Ghrelin modulates the activity and synaptic input organization of midbrain dopamine neurons while promoting appetite. J Clin Invest 116(12):3229–3239

Adan RA (2006) Constitutive receptor activity series: endogenous inverse agonists and constitutive receptor activity in the melanocortin system. Trends Pharmacol Sci 27(4):183–186

Arvanitakis L, Geras-Raaka E, Gershengorn MC (1998) Constitutively signaling G-protein-coupled receptors and human disease. Trends Endocrinol Metab 9(1):27–31

Bakker RA, Jongejan A, Sansuk K et al (2008) Constitutively active mutants of the histamine H1 receptor suggest a conserved hydrophobic asparagine-cage that constrains the activation of class A G protein-coupled receptors. Mol Pharmacol 73(1):94–103

Berson EL (1993) Retinitis pigmentosa. The friedenwald lecture. Invest Ophthalmol Vis Sci 34(5):1659–1676

Case R, Sharp E, Benned-Jensen T et al (2008) Functional analysis of the murine cytomegalovirus chemokine receptor homologue M33: ablation of constitutive signaling is associated with an attenuated phenotype in vivo. J Virol 82(4):1884–1898

Chow KB, Leung PK, Cheng CH et al (2008) The constitutive activity of ghrelin receptors is decreased by co-expression with vasoactive prostanoid receptors when over-expressed in human embryonic kidney 293 cells. Int J Biochem Cell Biol 40(11):2627–2637

Corvilain B, Van SJ, Dumont JE et al (2001) Somatic and germline mutations of the TSH receptor and thyroid diseases. Clin Endocrinol (Oxf) 55(2):143–158

Costa T, Cotecchia S (2005) Historical review: negative efficacy and the constitutive activity of G-protein-coupled receptors. Trends Pharmacol Sci 26(12):618–624

Daiger SP, Sullivan LS, Bowne SJ (2013) Genes and mutations causing retinitis pigmentosa. Clin Genet 84(2):132–141

Davenport AP, Bonner TI, Foord SM et al (2005) International union of pharmacology. LVI. Ghrelin receptor nomenclature, distribution, and function. Pharmacol Rev 57(4):541–546

De LA, Stadel JM, Lefkowitz RJ (1980) A ternary complex model explains the agonist-specific binding properties of the adenylate cyclase-coupled beta-adrenergic receptor. J Biol Chem 255(15):7108–7117

Deupi X, Kobilka BK (2010) Energy landscapes as a tool to integrate GPCR structure, dynamics, and function. Physiology (Bethesda) 25(5):293–303

Diano S, Farr SA, Benoit SC et al (2006) Ghrelin controls hippocampal spine synapse density and memory performance. Nat Neurosci 9(3):381–388

Elling CE, Frimurer TM, Gerlach LO et al (2006) Metal ion site engineering indicates a global toggle switch model for seven-transmembrane receptor activation. J Biol Chem 281(25):17337–17346

Gether U, Ballesteros JA, Seifert R et al (1997) Structural instability of a constitutively active G protein-coupled receptor. Agonist-independent activation due to conformational flexibility. J Biol Chem 272(5):2587–2590

Gjesing AP, Larsen LH, Torekov SS et al (2010) Family and population-based studies of variation within the ghrelin receptor locus in relation to measures of obesity. PLoS One 5(4):e10084

Hansen BS, Raun K, Nielsen KK et al (1999) Pharmacological characterisation of a new oral GH secretagogue, NN703. Eur J Endocrinol 141(2):180–189

Holliday ND, Holst B, Rodionova EA et al (2007) Importance of constitutive activity and arrestin-independent mechanisms for intracellular trafficking of the ghrelin receptor. Mol Endocrinol 21(12):3100–3112

Holst B, Cygankiewicz A, Jensen TH et al (2003) High constitutive signaling of the ghrelin receptor–identification of a potent inverse agonist. Mol Endocrinol 17(11):2201–2210

Holst B, Holliday ND, Bach A et al (2004) Common structural basis for constitutive activity of the ghrelin receptor family. J Biol Chem 279(51):53806–53817

Holst B, Lang M, Brandt E et al (2006) Ghrelin receptor inverse agonists: identification of an active peptide core and its interaction epitopes on the receptor. Mol Pharmacol 70(3):936–946

Holst B, Mokrosinski J, Lang M et al (2007a) Identification of an efficacy switch region in the ghrelin receptor responsible for interchange between agonism and inverse agonism. J Biol Chem 282(21):15799–15811

Holst B, Nygaard R, Valentin-Hansen L et al (2010) A conserved aromatic lock for the tryptophan rotameric switch in TM-VI of seven-transmembrane receptors. J Biol Chem 285(6):3973–3985

Holst B, Egerod K, Schwartz TW (2007b) Ghrelin: structural and functional properties. In: Sibley DR et al (eds) Handbook of Contemporary Neuropharmacology, vol 3., John Wiley & Sons IncHoboken, New Jersey, pp 765–783

Howard AD, Feighner SD, Cully DF et al (1996) A receptor in pituitary and hypothalamus that functions in growth hormone release. Science 273(5277):974–977

Jaeschke H, Neumann S, Kleinau G et al (2006) An aromatic environment in the vicinity of serine 281 is a structural requirement for thyrotropin receptor function. Endocrinology 147(4):1753–1760

Jensen AS, Sparre-Ulrich AH, Davis-Poynter N et al (2012) Structural diversity in conserved regions like the DRY-Motif among viral 7TM receptors-a consequence of evolutionary pressure? Adv Virol 2012:231813

Jerlhag E, Egecioglu E, Landgren S et al (2009) Requirement of central ghrelin signaling for alcohol reward. Proc Natl Acad Sci U S A 106(27):11318–11323

Jiang H, Betancourt L, Smith RG (2006) Ghrelin amplifies dopamine signaling by cross talk involving formation of growth hormone secretagogue receptor/dopamine receptor subtype 1 heterodimers. Mol Endocrinol 20(8):1772–1785

Katritch V, Cherezov V, Stevens RC (2013) Structure-function of the G protein-coupled receptor superfamily. Annu Rev Pharmacol Toxicol 53:531–556

Kim MS, Yoon CY, Park KH et al (2003) Changes in ghrelin and ghrelin receptor expression according to feeding status. NeuroReport 14(10):1317–1320

Kjelsberg MA, Cotecchia S, Ostrowski J et al (1992) Constitutive activation of the alpha 1B-adrenergic receptor by all amino acid substitutions at a single site. Evidence for a region which constrains receptor activation. J Biol Chem 267(3):1430–1433

Kojima M, Hosoda H, Date Y et al (1999) Ghrelin is a growth-hormone-releasing acylated peptide from stomach. Nature 402(6762):656–660

Kojima M, Hosoda H, Matsuo H et al (2001) Ghrelin: discovery of the natural endogenous ligand for the growth hormone secretagogue receptor. Trends Endocrinol Metab 12(3):118–122

Kudo M, Osuga Y, Kobilka BK et al (1996) Transmembrane regions V and VI of the human luteinizing hormone receptor are required for constitutive activation by a mutation in the third intracellular loop. J Biol Chem 271(37):22470–22478

Lau PN, Chow KB, Chan CB et al (2009) The constitutive activity of the ghrelin receptor attenuates apoptosis via a protein kinase C-dependent pathway. Mol Cell Endocrinol 299(2):232–239

Lefkowitz RJ, Cotecchia S, Samama P et al (1993) Constitutive activity of receptors coupled to guanine nucleotide regulatory proteins. Trends Pharmacol Sci 14(8):303–307

Lefkowitz RJ, Shenoy SK (2005) Transduction of receptor signals by beta-arrestins. Science 308(5721):512–517

Leung PK, Chow KB, Lau PN et al (2007) The truncated ghrelin receptor polypeptide (GHS-R1b) acts as a dominant-negative mutant of the ghrelin receptor. Cell Signal 19(5):1011–1022

Levin MC, Marullo S, Muntaner O et al (2002) The myocardium-protective Gly-49 variant of the beta 1-adrenergic receptor exhibits constitutive activity and increased desensitization and down-regulation. J Biol Chem 277(34):30429–30435

Liu G, Fortin JP, Beinborn M et al (2007) Four missense mutations in the ghrelin receptor result in distinct pharmacological abnormalities. J Pharmacol Exp Ther 322(3):1036–1043

Mary S, Fehrentz JA, Damian M et al (2013) Heterodimerization with its splice variant blocks the ghrelin receptor 1a in a nonsignaling conformation. A study with a purified heterodimer assembled into lipid discs. J Biol Chem 288(34):24656–24665

Matthews RP, Guthrie CR, Wailes LM et al (1994) Calcium/calmodulin-dependent protein kinase types II and IV differentially regulate CREB-dependent gene expression. Mol Cell Biol 14(9):6107–6116

Mokrosinski J, Frimurer TM, Sivertsen B et al (2012) Modulation of constitutive activity and signaling bias of the ghrelin receptor by conformational constraint in the second extracellular loop. J Biol Chem 287(40):33488–33502

Mokrosinski J, Holst B (2010) Modulation of the constitutive activity of the ghrelin receptor by use of pharmacological tools and mutagenesis. Methods Enzymol 484:53–73

Nogueiras R, Tovar S, Mitchell SE et al (2004) Regulation of growth hormone secretagogue receptor gene expression in the arcuate nuclei of the rat by leptin and ghrelin. Diabetes 53(10):2552–2558

Nygaard R, Frimurer TM, Holst B et al (2009) Ligand binding and micro-switches in 7TM receptor structures. Trends Pharmacol Sci 30(5):249–259

Okada M, Northup JK, Ozaki N et al (2004) Modification of human 5-HT(2C) receptor function by Cys23Ser, an abundant, naturally occurring amino-acid substitution. Mol Psychiatry 9(1):55–64

Pan Z, Lu T, Zhang X et al (2012) Identification of two mutations of the RHO gene in two Chinese families with retinitis pigmentosa: correlation between genotype and phenotype. Mol Vis 18:3013–3020

Pantel J, Legendre M, Cabrol S et al (2006) Loss of constitutive activity of the growth hormone secretagogue receptor in familial short stature. J Clin Invest 116(3):760–768

Perello M, Sakata I, Birnbaum S et al (2010) Ghrelin increases the rewarding value of high-fat diet in an orexin-dependent manner. Biol Psychiatry 67(9):880–886

Petersen PS, Woldbye DP, Madsen AN et al (2009) In vivo characterization of high Basal signaling from the ghrelin receptor. Endocrinology 150(11):4920–4930

Rasmussen SG, DeVree BT, Zou Y et al (2011) Crystal structure of the beta2 adrenergic receptor-Gs protein complex. Nature 477(7366):549–555

Rediger A, Piechowski CL, Habegger K et al (2012) MC4R dimerization in the paraventricular nucleus and GHSR/MC3R heterodimerization in the arcuate nucleus: is there relevance for body weight regulation? Neuroendocrinology 95(4):277–288

Rediger A, Piechowski CL, Yi CX et al (2011) Mutually opposite signal modulation by hypothalamic heterodimerization of ghrelin and melanocortin-3 receptors. J Biol Chem 286(45):39623–39631

Ren Q, Kurose H, Lefkowitz RJ et al (1993) Constitutively active mutants of the alpha 2-adrenergic receptor. J Biol Chem 268(22):16483–16487

Rivera-De la Parra D, Cabral-Macias J, Matias-Florentino M et al (2013) Rhodopsin p. N78I dominant mutation causing sectorial retinitis pigmentosa in a pedigree with intrafamilial clinical heterogeneity. Gene 519(1):173–176

Rovati GE, Capra V, Neubig RR (2007) The highly conserved DRY motif of class A G protein-coupled receptors: beyond the ground state. Mol Pharmacol 71(4):959–964

Samama P, Cotecchia S, Costa T et al (1993) A mutation-induced activated state of the beta 2-adrenergic receptor. Extending the ternary complex model. J Biol Chem 268(7):4625–4636

Scheerer P, Park JH, Hildebrand PW et al (2008) Crystal structure of opsin in its G-protein-interacting conformation. Nature 455(7212):497–502

Schellekens H, van Oeffelen WE, Dinan TG et al (2013) Promiscuous dimerization of the growth hormone secretagogue receptor (GHS-R1a) attenuates ghrelin-mediated signaling. J Biol Chem 288(1):181–191

Schneider EH, Schnell D, Strasser A et al (2010) Impact of the DRY motif and the missing "ionic lock" on constitutive activity and G-protein coupling of the human histamine H4 receptor. J Pharmacol Exp Ther 333(2):382–392

Schwartz TW, Frimurer TM, Holst B et al (2006) Molecular mechanism of 7TM receptor activation–a global toggle switch model. Annu Rev Pharmacol Toxicol 46:481–519

Seifert R, Wenzel-Seifert K (2002) Constitutive activity of G-protein-coupled receptors: cause of disease and common property of wild-type receptors. Naunyn Schmiedebergs Arch Pharmacol 366(5):381–416

Singh LP, Andy J, Anyamale V et al (2001) Hexosamine-induced fibronectin protein synthesis in mesangial cells is associated with increases in cAMP responsive element binding (CREB) phosphorylation and nuclear CREB: the involvement of protein kinases A and C. Diabetes 50(10):2355–2362

Sivertsen B, Holliday N, Madsen AN et al (2013) Functionally biased signalling properties of 7TM receptors—opportunities for drug development for the ghrelin receptor. Br J Pharmacol 170(7):1349–1362

Sivertsen B, Lang M, Frimurer TM et al (2011) Unique interaction pattern for a functionally biased ghrelin receptor agonist. J Biol Chem 286(23):20845–20860

Smit MJ, Vischer HF, Bakker RA et al (2007) Pharmacogenomic and structural analysis of constitutive g protein-coupled receptor activity. Annu Rev Pharmacol Toxicol 47:53–87

Steen A, Thiele S, Guo D et al (2013) Biased and constitutive signaling in the CC-chemokine receptor CCR5 by manipulating the interface between transmembrane helices 6 and 7. J Biol Chem 288(18):12511–12521

Tolle V, Low MJ (2008) In vivo evidence for inverse agonism of Agouti-related peptide in the central nervous system of proopiomelanocortin-deficient mice. Diabetes 57(1):86–94

Vishnivetskiy SA, Ostermaier MK, Singhal A et al (2013) Constitutively active rhodopsin mutants causing night blindness are effectively phosphorylated by GRKs but differ in arrestin-1 binding. Cell Signal 25(11):2155–2162

White JF, Noinaj N, Shibata Y et al (2012) Structure of the agonist-bound neurotensin receptor. Nature 490(7421):508–513

Yanagawa M, Yamashita T, Shichida Y (2013) Glutamate acts as a partial inverse agonist to metabotropic glutamate receptor with a single amino acid mutation in the transmembrane domain. J Biol Chem 288(14):9593–9601

Homodimerization and Heterodimerization of the Ghrelin Receptor

Alessandro Laviano and Alessia Mari

Abstract Ghrelin triggers different metabolic and behavioral effects. This suggests that the GHSR may act synergistically with other families of receptors. Two isoforms of GHSR have been identified, i.e., GHSR-1a and GHSR-1b. Only GHSR-1a is the active form and transduces ghrelin signal. However, being part of GPCR family, GHSR-1a may form homo- and heterodimers. Consistent evidence shows that GHSR-1a/GHSR-1b heterodimers reduce the intracellular signaling triggered by ghrelin. Also, heterodimers consisting of GHSR-1a and DA, 5-HT and MC3R have been also described and functionally characterized, and provide mechanistic explanation of the impact of the ghrelin system on different neuronal pathways.

Keywords Ghrelin · GHSR · Ghrelin receptor 1a · Ghrelin receptor 1b · Homodimer · Heterodimer · 5-HT · DA · MC3R

Introduction

A key factor favoring survival and evolution of living organisms is the ability to efficiently control energy metabolism. In particular, nutrient availability may greatly vary according to seasonal or environmental changes. Therefore, the development of metabolic pathways preserving cell mass during famine and storing excess energy during feast represented a critical achievement in the evolution of life on planet Earth. Considering the importance of preserving energy homeostasis, a number of regulatory pathways have evolved and redundant pathways still coexist in animals and humans. Among them, the ghrelin/GOAT/

A. Laviano (✉) · A. Mari
Department of Clinical Medicine, Sapienza University, Viale del Policlinico 155,
00161 Rome, Italy
e-mail: alessandro.laviano@uniroma1.it

J. Portelli and I. Smolders (eds.), *Central Functions of the Ghrelin Receptor*,
The Receptors 25, DOI: 10.1007/978-1-4939-0823-3_2,
© Springer Science+Business Media New York 2014

ghrelin receptor system appears to play a biologically and clinically relevant role during health and disease. In fact, this system contributes to the control of a number of key pathways including energy metabolism, GH secretion, inflammatory response, glucose metabolism, cardiovascular performance, and behavior (Lim et al. 2011).

Structure and Distribution of the Ghrelin Receptor (Ghrelin Receptor 1a and Ghrelin Receptor 1b)

The natural ligand of the ghrelin receptor is acyl ghrelin. However, ghrelin receptor identification in 1996 preceded that of ghrelin, which was isolated 3 years later (Kojima et al. 1999). Initially, the biological effects of ghrelin receptor were investigated by using synthetic peptides, which contributed to link ghrelin receptor activity to the potent induction of GH secretion. Later, when ghrelin became available, studies revealed that ghrelin and its receptor also influence food intake, gut motility, sleep, memory and behavior, glucose and lipid metabolism, cardiovascular performances, cell proliferation, immunological responses, reproduction, and cell apoptosis (Peter et al. 2008; Lau et al. 2009). These results provided evidence of the complexity of the ghrelin/GOAT/GHSR system and suggested that GHSR receptor distribution extended well beyond the boundaries of the central nervous system.

Indeed, GHSR is localized both in non-nervous organs/tissues (i.e., adipose tissue, myocardium, adrenals, gonads, lung, liver, arteries, stomach, pancreas, thyroid, and kidney) as well as in CNS and higher levels of expression in the pituitary gland and the hypothalamus and lower levels of expression in other organs have been reported (Lattuada et al. 2013). Since the characterization of GHSR within the central nervous system is of extreme importance in order to gain insights on its role in the pathogenesis of clinically relevant neurodegenerative events, it has been recently demonstrated that GHSR is expressed in primary neurons and that its expression is dependent upon their developmental stage (Lattuada et al. 2013). Moreover, GHSR expression shows differences according to the brain region involved, with a more pronounced expression in hippocampal rather than cortical neurons. Supporting the importance of tissue development in modulating GHSR expression, Wang et al. have recently shown that the number of ghrelin-immune positive cells increases with age in the African ostrich GI tract from postnatal day 1 to day 90, which suggests that GHSR may be involved in GI tract development (Wang et al. 2009).

The GHSR is comprised within the family of GPCR, and is characterized by a seven transmembrane domain protein consisting of 366 amino acids. GHSR is linked to G(q) and G(s) signaling pathways, and the binding of ghrelin or synthetic peptidyl and nonpeptidyl agonists leads to increased intracellular Ca^{2+} content. The molecular mechanisms by which GHSR mediates biological functions are complex, and involve intracellular signaling pathways which are specific of the tissue type in which GHSR is expressed (Soares et al. 2008). Moreover, GHSR shares with other GPCRs the ability to form homodimers and heterodimers which results in the formation of receptor complexes with altered trafficking, signaling, and pharmacological properties.

The gene encoding GHSR has been detected on chromosome 3q26.2 (McKee et al. 1997). Interestingly, sequence homologies have been identified with the motilin receptor, with approximately 40 % sequence identity, and the neurotensin receptor (Feighner et al. 1999). Recent data show that the GHSR and motilin receptor (GPR38) shares not only large sequence identity, but also tissue distribution (Suzuki et al. 2012). GHSR mRNA expressions have been detected throughout the stomach and intestine, whereas GPR38 has been found to be expressed in the gastric muscle layer, lower intestine, lungs, heart, and pituitary gland (Suzuki et al. 2012). These results suggest that gut motility and energy metabolism are closely related and controlled by specific receptors expressed in the gastrointestinal tract and/or in the CNS.

The GHSR gene encodes for the functional ghrelin receptor. Two isoforms of the GHSR have been identified, i.e., GHSR-1a and GHSR-1b. Only GHSR-1a transduces ghrelin's signal by binding the active form of ghrelin, i.e., ghrelin with GOAT-mediated O-n-octanoyl acid modification at serine 3 position (acyl ghrelin) (Gomez et al. 2009). GHSR-1a is a constitutively active GPCR and is mainly expressed in the pituitary and at a lower level in a number of hypothalamic nuclei, particularly the ARC, the VMN and the PVN of the hypothalamus (Gnanapavan et al. 2002; Guan et al. 1997). However, other brain areas also express the GHSR-1a, including the substantia nigra, the dorsal and median raphe nuclei, the ventral tegmental area and the hippocampus (Guan et al. 1997). GHSR-1a is also expressed in peripheral tissues: pancreas, spleen, myocardium, adrenal gland, adipose tissue, intestine and blood vessels (Gnanapavan et al. 2002; Schellekens et al. 2010). Recently, a role for GHSR-1a in the kidney has been also identified. GHSR-1a expression has been found in the straight parts of the distal tubules and the thin limbs of the loops of Henle (Venables et al. 2011). No expression was detected in other structures, including the glomeruli, proximal tubules and collecting ducts (Venables et al. 2011). GHSR-1a was not found in extra-renal or intra-renal arteries, despite observations that ghrelin is a vasodilator (Venables et al. 2011). Therefore, it seems that GHSR-1a has a restricted distribution in the kidney and possibly mediates sodium retention.

A more functional analysis of GHSR-1a tissue distribution has been obtained by assessing plasma activity of exogenous ghrelin and its distribution in rats. Ruchala et al. measured plasma radioactivity of ^{125}I-ghrelin in blood and tissue specimens collected after ^{125}I-ghrelin administration (Ruchala et al. 2012) Plasma ^{125}I-ghrelin radioactivity decreased rapidly after peptide administration. The half-life time of ^{125}I-ghrelin was 15–18 min (Ruchala et al. 2012). The analysis of ^{125}I-ghrelin distribution revealed three profiles of tissue uptake. The first profile was characterized by decreasing radioactivity (i.e., brain, kidney, liver) (Ruchala et al. 2012). Increasing tissue radioactivity followed by a gradual decrease (second profile) was observed in stomach, intestine and thyroid (Ruchala et al. 2012). The third profile was described as a relatively stable radioactivity (i.e., lung, myocardium) (Ruchala et al. 2012). When considered together, these results indicate that the functional activities mediated by GHSR-1a are numerous and extend beyond the mere regulation of GH secretion.

As previously mentioned, the molecular intracellular mechanisms transducing the ghrelin signal are complex. Stimulation of GHSR-1a by GH secretagogues evokes increases in intracellular Ca^{2+} concentration, whereas GHSR-1b appears to play an inhibitory role on the signal transduction activity of GHSR-1a (Chan et al. 2004). By using GH secretagogues, namely, GHRP-6 and L163,540, Chan et al. were able to show that these ligands trigger a receptor specific and phospholipase C (PLC)-dependent elevation of intracellular Ca^{2+} in HEK293 cells stably expressing ghrelin receptor 1a (Chan et al. 2004). This GH secretagogue-induced Ca^{2+} mobilization is also dependent on protein kinase C activated L-type Ca^{2+} channel opening. Also, It was found that ghrelin receptor 1a could function in an agonist-independent manner as it exhibited a high basal activity of IP_3 production in the absence of GH secretagogues, indicating that the receptor is constitutively active (Chan et al. 2004). In addition, the extracellular signal-regulated kinases 1 and 2 (ERK1/2) were found to be activated upon stimulation of ghrelin receptor 1a by GHRP-6 (Chan et al. 2004). Neither Gs nor Gi proteins are coupled to the receptor, as GH secretagogues did not induce cAMP production nor inhibit forskolin-stimulated cAMP accumulation in the ghrelin receptor 1a bearing cells. Thus, ghrelin receptor 1a appears to couple through the G(q/11)-mediated pathway to activate PLC, resulting in increased IP_3 production and Ca^{2+} mobilization from both intracellular and extracellular stores. Moreover, ghrelin receptor 1a may trigger multiple signal transduction cascades to exert its physiological functions.

Ghrelin receptor 1b is a truncated receptor variant with only five transmembrane domains and consisting of 289 amino acids. This variant is not a singularity within the family of G protein coupled hormone receptors. In fact, many other truncated variants have been identified, including the α1A-adrenergic receptor, the dopamine D3 receptor, the gonadotrophin receptor and the V2 vasopressin receptor (Leung et al. 2007). The biological function of ghrelin receptor 1b, which does not bind ghrelin nor other GH secretagogues, is being elucidated. However, ghrelin receptor 1b has a larger tissue distribution than ghrelin receptor 1a (Gnanapavan et al. 2002), which suggests a relevant functional role for this inactive variant. Indeed, consistent evidence suggests that ghrelin receptor 1b acts as a dominant-negative mutant of ghrelin receptor 1a, thus negatively influencing ghrelin receptor 1a function (Leung et al. 2007).

Accumulating evidence, as reviewed in (Muccioli et al. 2007), indicates that ghrelin receptor 1a may not be the only responsible receptor for all the effects mediated by ghrelin. As an example, ghrelin receptor 1a-deficient mice are similar to wild type animals in growth and diet-induced obesity (DIO), whereas ghrelin and the nonacylated form of ghrelin (des-acyl ghrelin), which does not bind ghrelin receptor 1a, share the same biological actions on the heart, adipose tissue, pancreas, cancer cells and brain. These results suggest the existence of a still unknown, functionally active binding site for this family of molecules. Interestingly, a number of variants of ghrelin receptor 1a and ghrelin receptor 1b have been identified. Kaiya et al. identified cDNA that encodes protein with close sequence similarity to ghrelin receptor and exon–intron organization of the ghrelin receptor genes in rainbow trout (Kaiya et al. 2009a). Two variants of the ghrelin receptor 1a proteins with 387-amino acids, namely DQTA/LN-type and ERAT/IS-type, were identified (Kaiya et al. 2009a). In $3'$-RACE PCR and genomic PCR, three ghrelin receptor 1b orthologs were identified, which consisted of 297- or 300-amino acids with different amino acid sequence at the C-terminus, in addition to the DQTA/LN-type and ERAT/IS-type variations. Genomic PCR revealed that the genes are composed of two exons separated by an intron, and that two ghrelin receptor 1a and three ghrelin receptor 1b variants are generated by three distinct genes (Kaiya et al. 2009a). Identified DQTA/LN-type or ERAT/IS-type ghrelin receptor 1a cDNA was transfected into mammalian cells, and intracellular Ca^{2+} ion mobilization assay was carried out (Kaiya et al. 2009a). However, no response to rat ghrelin nor to a homologous ligand of either receptor in vitro was found. Similarly, a ghrelin receptor-like receptor was identified in the Mozambique tilapia (Kaiya et al. 2009b). Although gene structures and characterization of protein sequences identified in these studies were closely similar to other ghrelin receptor, further studies are required to conclude that they are species-specific ghrelin receptor.

Ghrelin Receptor Dimers

GPCRs represent the largest group of cell surface receptors and an important pharmacological target. Though originally thought to act in a one receptor-one effector fashion, it is now known that these receptors are capable of oligomerization and can function as dimers or higher order oligomers in native tissue (Wertman and Dupré 2013). They do not only assemble with identical receptors as homodimers, but also associate with different GPCRs to form heterodimers. Interestingly, GPCRs homo- and heterodimers are regulated by different chaperones, Rabs, and scaffolding proteins, further emphasizing their potential as unique targets. Ghrelin receptor dimerization plays a significant role not only in protein trafficking and expression on cell surface, but it also impacts on intracellular signaling. However, from a physiological point of view, ghrelin receptor 1a dimerization to form homodimers is less appealing since heterodimers appear to exert more complex molecular and metabolic effects. Therefore, ghrelin receptor heterodimers received more attention by researchers, since they may explain the functional relationship of the ghrelin/GOAT/ghrelin receptor system with other metabolic pathways.

GHS Heterodimers

Ghrelin Receptor 1a and Ghrelin Receptor 1b

Ghrelin receptor 1a is a constitutively active (i.e., agonist-independent) receptor. However, fine tuning of its efficiency to transduce the signal triggered by ghrelin is critical to effectively adapt to different metabolic challenges. Therefore, intracellular molecular events should occur to increase or decrease the activity of ghrelin receptor 1a (Table 1).

It is well established that truncated GPCRs, i.e., those arising through mRNA splicing, may modulate the function of the full-length version of the receptor by physical interaction. Ghrelin receptor 1b is no exception. Leung et al. demonstrated that ghrelin receptor 1a and ghrelin receptor 1b can exist as heterodimers, and that the formation of heterodimers prevents agonist-dependent changes in the energy transfer seen with dimers of ghrelin receptor 1a (Leung et al. 2007). Furthermore, an excess of ghrelin receptor 1b mRNA compared with ghrelin receptor 1a mRNA attenuates trafficking of ghrelin receptor 1a to the cell surface and decreases constitutive activity, yet does not diminish ghrelin-stimulated intracellular signaling (Leung et al. 2007). Confirming these results, it has been recently shown that seabream ghrelin receptor 1b can inhibit ghrelin receptor 1a-stimulated Ca^{2+} mobilization in human embryonic cells stably expressing seabream ghrelin receptor 1a (Chan and Cheng 2004). Interestingly, ghrelin receptor 1b has an intracellular localization distinct from ghrelin receptor 1a, being primarily localized in the endoplasmic reticulum (Chow et al. 2012). Immunocytochemical studies suggest that ghrelin receptor 1b decreases the plasma membrane expression of ghrelin receptor 1a, but the overall distribution profile of ghrelin receptor 1a in isolated subcellular fractions is unaffected by ghrelin receptor 1b. Using bioluminescence resonance energy transfer methods, it has been shown that while ghrelin receptor 1a dimers are evenly distributed in all subcellular fractions, ghrelin receptor 1a/ghrelin receptor 1b heterodimers are concentrated within the endoplasmic reticulum (Chow et al. 2012). These results suggest that ghrelin receptor 1b traps ghrelin receptor 1a within the endoplasmic reticulum by the process of oligomerization. Furthermore, ghrelin receptor 1a constitutively activated ERK1/2 in the endoplasmic reticulum, but this small response was not affected by ghrelin receptor 1b and its physiological relevance is uncertain (Chow et al. 2012). Taken together, these results suggest that ghrelin receptor 1a can be retained in the endoplasmic reticulum by heterodimerization with ghrelin receptor 1b, and constitutive activation of phospholipase C is attenuated due to decreased cell surface expression of ghrelin receptor 1a. However, sufficient ghrelin receptor 1a homodimers can still be expressed on the cell surface for maximal responses to agonist stimulation.

As previously mentioned, heterodimerization of GPCRs has an impact on their signaling properties, but the molecular mechanisms underlying heteromer-directed selectivity remain elusive. Using purified monomers and dimers reconstituted into

Table 1 Isoforms of GHSR, and family of receptors interacting with GHSR-1a

	Structure	Function	Mechanism(s) of function
GHSR-1a	GPCR (7 transmembrane domains)	Transduces ghrelin's signal	Constitutionally active receptor, increasing intracellular Ca^{2+}
GHSR-1b	GPCR-truncated variant (5 transmembrane domains)	Inhibits/modulates activity of GHSR-1a	Heterodimerization with GHSR-1a, yielding reduced expression of GHSR-1a on cell surface
DAD1	GPCR (7 transmembrane domains)	Dopaminergic neurotransmission	Heterodimerization with GHSR-1a amplifies DA signaling
MC3R	GPCR (7 transmembrane domains)	Melanocortin signaling	Heterodimerization with GHSR-1a attenuates GHSR-1a signaling
$5\text{-}HT_{2c}$	GPCR (7 transmembrane domains)	Serotonergic neurotransmission	Heterodimerization with GHSR-1a attenuates GHSR-1a signaling

lipid disks, Mary et al. explored whether an alternative strategy could be used by ghrelin receptor 1b to reduce ghrelin receptor 1a activity, beyond retention of ghrelin receptor 1a in the endoplasmic reticulum (Mary et al. 2013). Their research hypothesis was to investigate how dimerization impacts on the functional and structural behavior of ghrelin receptor 1a. In particular, they studied how a naturally occurring truncated splice variant of ghrelin receptor 1a receptor exerts a dominant negative effect on ghrelin signaling upon dimerization with the full-length receptor. Results obtained provide direct evidence that this dominant negative effect is due to the ability of the nonsignaling truncated receptor to restrict the conformational landscape of the full-length protein (Mary et al. 2013). Indeed, associating both proteins within the same disk blocks all agonist- and signaling protein-induced changes in GHSR-1a conformation, thus preventing it from activating its cognate G protein and triggering arrestin-2 recruitment. This is an unambiguous demonstration that allosteric conformational events within dimeric assemblies can be directly responsible for modulation of signaling mediated by GPCRs (Mary et al. 2013).

DA, MC3R, 5-HT

The ghrelin/GOAT/ghrelin receptor system is involved in mediating biological effects, which appear independent and not strictly related to energy homeostasis, i.e., tissue development or gut motility. On the other hand, it contributes to appetite control, which is regulated by many central and peripheral neuronal pathways.

This evidence highlights the close functional interaction existing between ghrelin receptor and other receptor families activating independent neuronal output. Such functional relationship could be secondary to anatomical co-localization of ghrelin receptor and other receptors, as it is the case of ghrelin receptor and GPR38. However, it is tempting to speculate that in neurons co-expressing ghrelin receptor and other receptor families, a physical and functional interaction may occur. In particular, considering the ability of GPCRs to form heterodimers, the possibility that ghrelin receptor may physically interact with neurotransmitter or neuropeptide receptors has been extensively explored.

Ghrelin has been consistently demonstrated to modulate neuronal activity in the brain. To identify neurons that express ghrelin receptor, ghrelin receptor-IRES-tauGFP mice were generated by gene targeting (Jiang et al. 2006). Neurons expressing the ghrelin receptor exhibit green fluorescence and are clearly evident in the hypothalamus, hippocampus, cortex, and midbrain. Using immunohisto-chemistry in combination with green fluorescent protein fluorescence, neurons that co-express DAD1 and ghrelin receptor 1a were identified. Further experiments in vitro showed that activation of ghrelin receptor 1a by ghrelin amplifies DA/DAD1-induced cAMP accumulation (Jiang et al. 2006). Intriguingly, amplification involves a switch in G protein coupling of the ghrelin receptor 1a from Galpha(11/q) to Galpha(i/o) by a mechanism consistent with agonist-dependent formation of ghrelin receptor 1a/DAD1 heterodimers (Jiang et al. 2006). These results indicate that ghrelin has the potential to amplify DA signaling selectively in neurons that co-express DAD1 and ghrelin receptor 1a.

Food intake is a complex behavior which is tightly and redundantly regulated in the hypothalamus. Impairment of the physiological interactions occurring between ghrelin receptor 1a, serotonergic neurotransmission and melanocortin signaling result in cachexia (Laviano et al. 2008) or obesity (Nonogaki et al. 2006). Inter-estingly, the ghrelin receptor 1a, MC3R, and the serotonin 2C receptor (5-HT$_{2C}$) are all GPCRs. Schellekens et al. investigated the downstream signaling conse-quences and ligand-mediated co-internalization following heterodimerization of the ghrelin receptor 1a receptor with DAD1, as well as that of the ghrelin receptor 1a-MC3R heterodimer (Schellekens et al. 2013). In addition, a novel heterodimer between the ghrelin receptor 1a receptor and the 5-HT$_{2C}$ was identified (Schellekens et al. 2013). Interestingly, dimerization of ghrelin receptor 1a with the unedited 5-HT$_{2C}$-INI, but not with the partially edited 5-HT$_{2C}$-VSV isoform, significantly reduced ghrelin receptor 1a agonist-mediated Ca^{2+} influx, which was completely restored following pharmacological blockade of 5-HT$_{2C}$ (Schellekens et al. 2013). These results suggest a potential novel mechanism for fine-tuning ghrelin receptor 1a receptor-mediated activity via promiscuous dimerization of the ghrelin receptor 1a receptor with other GPCRs involved in appetite regulation and food reward. These findings may uncover novel mechanisms of significant rele-vance for the future pharmacological targeting of the ghrelin receptor 1a receptor in the homeostatic regulation of energy balance and in hedonic appetite signaling, both of which play a significant role in the development of obesity.

Emerging evidence suggests that ghrelin is a directly acting vasodilator peptide with anti-inflammatory activity. Therefore, the ability of ghrelin receptor 1a to oligomerize with members of the prostanoid receptor family, which are also involved in modulating vascular activity and inflammatory responses, has been explored. Using the techniques of bioluminescence resonance energy transfer and co-immunoprecipitation, ghrelin receptor 1a has been demonstrated to be able to hetero-oligomerize with prostaglandin E2 receptor subtype (EP_{3-I}), prostacyclin receptors, and thromboxane A_2 ($TP\alpha$) receptors, when transiently over-expressed in human embryonic kidney 293 cells (Chow et al. 2008). These results suggest that hetero-oligomeric interactions between ghrelin receptor 1a and prostanoid receptors are likely to be of biological relevance. Co-transfection of cells with ghrelin receptor 1a and prostanoid receptors significantly decreased ghrelin receptor 1a expression and attenuated its constitutive activation of PLC without changing its affinity for ghrelin (Chow et al. 2008). An increase in the proportion of ghrelin receptor 1a localized intracellularly in the presence of prostanoid receptors has been also observed (Chow et al. 2008). Taken together, these results suggest that the increased expression of prostanoid receptors in conditions of vascular inflammation, such as in atherosclerotic plaques, could influence those cellular responses dependent on the constitutive activation of ghrelin receptor 1a.

Conclusions

The ghrelin/GOAT/ghrelin receptor system plays a key role in fine-tuning human metabolism and in precisely adapting energy homeostasis with environmental and developmental challenges. Consequently, the functional impairment of this system contributes to the onset of diseases, including disease-associated malnutrition and obesity (Pantel et al. 2006). The better understanding of the physiology of ghrelin receptor may help in developing effective therapeutic strategies for those diseases characterized by the failure of energy homeostasis. In this light, the discovery that heterodimerization of ghrelin receptor 1a results in profound functional consequences promises new avenues for investigation and understanding of hypothalamic functions dependent on GPCR signaling (Rediger et al. 2012). In fact, since GPCRs are important targets for drugs to combat many diseases, identification of heterodimers may be a prerequisite for highly specific drugs.

References

Chan CB, Leung PK, Wise H, Cheng CH (2004) Signal transduction mechanism of the seabream growth hormone secretagogue receptor. FEBS Lett 577:147–153
Chan CB, Cheng CH (2004) Identification and functional characterization of two alternatively spliced growth hormone secretagogue receptor transcripts from the pituitary of black seabream Acanthopagrus schlegeli. Mol Cell Endocrinol 214:81–95

Chow KB, Leung PK, Cheng CHK, Cheung WT. Wise H (2008) The constitutive activity of ghrelin receptors is decreased by co-expression with vasoactive prostanoid receptors when over-expressed in human embryonic kidney 293 cells. Int J Biochem Cell Biol 40:2627–2637

Chow KB, Sun J, Chu KM, Tai Cheung W, Cheng CH, Wise H (2012) The truncated ghrelin receptor polypeptide (GHSR-1b) is localized in the endoplasmic reticulum where it forms heterodimers with ghrelin receptors (GHSR-1a) to attenuate their cell surface expression. Mol Cell Endocrinol 348:247–354

Feighner SD, Tan CP, McKee KK et al (1999) Receptor for motilin identified in the human gastrointestinal system. Science 284:2184–2188

Gnanapavan S, Kola B, Bustin SA et al (2002) The tissue distribution of the mRNA of ghrelin and subtypes of its receptor, GHS-R, in humans. J Clin Endocrinol Metab 87:2988–2991

Gomez R, Lago F, Gomez-Reino JJ, Gualillo O (2009) Novel factors as therapeutic targets to treat diabetes. Focus on leptin and ghrelin. Expert Opin Ther Targets 13:583–591

Guan XM, Yu H, Palyha OC et al (1997) Distribution of mRNA encoding the growth hormone secretagogue receptor in brain and peripheral tissues. Brain Res Mol Brain Res 48:23–29

Jiang H, Betancourt L, Smith RG (2006) Ghrelin amplifies dopamine signaling by cross talk involving formation of growth hormone secretagogue receptor/dopamine receptor subtype 1 heterodimers. Mol Endocrinol 20:1772–1785

Kaiya H, Mori T, Miyazato M, Kangawa K (2009a) Ghrelin receptor (GHS-R)-like receptor and its genomic organisation in rainbow trout, Oncorhynchus mykiss. Comp Biochem Physiol A Mol Integr Physiol 153:438–509

Kaiya H, Riley LG, Janzen W, Hirano T, Grau EG, Miyazato M et al (2009b) Identification and genomic sequence of a ghrelin receptor (GHS-R)-like receptor in the Mozambique tilapia, Oreochromis mossambicus. Zoolog Sci 26:330–337

Kojima M, Hosoda H, Date Y, Nakazato M, Matsuo H, Kangawa K (1999) Ghrelin is a growth-hormone-releasing acylated peptide from stomach. Nature 402:656–660

Lattuada D, Crotta K, Tonna N et al (2013) The expression of GHS-R in primary neurons is dependent upon maturation stage and regional localization. PLoS ONE 8(6):e64183

Lau PN, Chow KB, Chan CB, Cheng CH, Wise H (2009) The constitutive activity of the ghrelin receptor attenuates apoptosis via a protein kinase C-dependent pathway. Mol Cell Endocrinol 299:232–239

Laviano A, Inui A, Marks DL et al (2008) Neural control of the anorexia-cachexia syndrome. Am J Physiol Endocrinol Metab 295:E1000–E1008

Leung PK, Chow KB, Lau PN et al (2007) The truncated ghrelin receptor polypeptide (GHSR-1b) acts as a dominant-negative mutant of the ghrelin receptor. Cell Signal 19:1011–1022

Lim CT, Kola B, Korbonits M (2011) The ghrelin/GOAT/GHS-R system and energy metabolism. Rev Endocr Metab Disord 12:173–186

Mary S, Fehrentz JA, Damian M, Gaibelet G, Orcel H, Verdié P et al (2013) Heterodimerization with its splice variant blocks the ghrelin receptor 1a in a nonsignaling conformation. A study with a purified heterodimer assembled into lipid discs. J Biol Chem Jul 9 (Epub ahead of print)

McKee KK, Palyha OC, Feighner SD et al (1997) Molecular analysis of rat pituitary and hypothalamic growth hormone secretagogue receptors. Mol Endocrinol 11:415–423

Muccioli G, Baragli A, Granata R, Papotti M, Ghigo E (2007) Heterogeneity of ghrelin/growth hormone secretagogue receptors. Toward the understanding of the molecular identity of novel ghrelin/GHS receptors. Neuroendocrinology 86:147–164

Nonogaki K, Nozue K, Oka Y (2006) Hyperphagia alters expression of hypothalamic 5-HT2C and 5-HT1B receptor genes and plasma des-acyl ghrelin levels in Ay mice. Endocrinology 147:5893–5900

Pantel J, Legendre M, Cabrol S et al (2006) Loss of constitutive activity of the growth hormone secretagogue receptor in familial short stature. J Clin Invest 116:760–768

Peter P, Beatrix S, Eva R et al (2008) Ghrelin: a new peptide regulating the neurohormonal system, energy homeostasis and glucose metabolism. Diabetes Metab Res Rev 24:343–352

Rediger A, Piechowski CL, Habegger K et al (2012) MC4R dimerization in the paraventricular nucleus and GHSR/MC3R heterodimerization in the arcuate nucleus: is there relevance for body weight regulation? Neuroendocrinology 95:277–288

Ruchala M, Rafinska L, Kosowicz J et al (2012) The analysis of exogenous ghrelin plasma activity and tissue distribution. Neuro Endocrinol Lett 33:191–195

Schellekens H, Dinan TG, Cryan JF (2010) Lean mean fat reducing "ghrelin" machine: Hypothalamic ghrelin and ghrelin receptors as therapeutic targets in obesity. Neuropharmacology 58:2–16

Schellekens H, van Oeffelen WE, Dinan TG, Cryan JF (2013) Promiscuous dimerization of the growth hormone secretagogue receptor (GHSR-1a) attenuates ghrelin-mediated signaling. J Biol Chem 288:181–191

Soares JB, Roncon-Albuquerque R, Leite-Moreira A (2008) Ghrelin and ghrelin receptor inhibitors: agents in the treatment of obesity. Expert Opin Ther Targets 12:1177–1189

Suzuki A, Ishida Y, Aizawa S et al (2012) Molecular identification of GHS-R and GPR38 in Suncus murinus. Peptides 36:29–38

Venables G, Hunne B, Bron R, Cho HJ, Brock JA, Furness JB (2011) Ghrelin receptors are expressed by distal tubules of the mouse kidney. Cell Tissue Res 346:135–139

Wang JX, Peng KM, Liu HZ, Song H, Chen X, Liu M (2009) Distribution and developmental changes in ghrelin- immunopositive cells in the gastrointestinal tract of African ostrich chicks. Regul Pept 154:97–101

Wertman J, Dupré DJ (2013) G protein-coupled receptor dimers: look like their parents, but act like teenagers! J Recept Signal Transduct Res 33:135–138

Part II

Mechanisms of Ghrelin's Action

The Role of the Ghrelin Receptor in Appetite and Energy Metabolism

Romana Stark and Zane B. Andrews

Abstract Ghrelin is a stomach hormone secreted into the bloodstream that acts on ghrelin receptors (GHSR1a) in the hypothalamus to increase food intake and regulate energy metabolism. This review focuses on the role of the GHSR1a in the hypothalamus and highlights the function the different nuclei expressing the GHSR1a. We discuss the mechanisms through which ghrelin activates receptors on NPY neurons and downstream signaling within NPY neurons. The downstream signaling involves a number of key metabolic signaling nodes including CaMKK, AMPK, CPT1, UCP2 and SIRT1 pathways that enhances mitochondrial efficiency and buffers reactive oxygen species in order to maintain an appropriate firing response in NPY. Finally, we examine a new model of synaptic plasticity in hypothalamic feeding circuits in which ghrelin activates GHSR1a on presynaptic glutamatergic inputs onto NPY and switches on an AMPK-dependent feed-forward system. This model of synaptic plasticity ensures sustained NPY firing during periods of negative energy balance. Taken together, we detail a number of novel mechanisms through which ghrelin signaling via the GHSR1a maintains high NPY neuronal activity in order to promote food intake under conditions of negative energy balance.

Keywords Neuropeptide Y (NPY) · Agouti-related peptide (AgRP) · Hypothalamus · AMP-activated kinase (AMPK) · Ghrelin · Mitochondria · Arcuate nucleus

R. Stark · Z. B. Andrews (✉)
Department of Physiology, Monash University, Clayton, VIC 3183, Australia
e-mail: Zane.andrews@monash.edu

J. Portelli and I. Smolders (eds.), *Central Functions of the Ghrelin Receptor*,
The Receptors 25, DOI: 10.1007/978-1-4939-0823-3_3,
© Springer Science+Business Media New York 2014

Introduction

The brain plays a critical role in the regulation of appetite, body weight, and energy homeostasis, and recent genome wide association studies show that human obesity is largely a heritable disorder affecting the neural control of energy balance (Farooqi and O'Rahilly 2006; Loos et al. 2008; O'Rahilly and Farooqi 2008; Willer et al. 2009). In order to serve an important role in energy homeostasis, the central nervous system (CNS) receives feedback information from peripheral tissues in the form of hormones, nutrients, or afferent sensory neural information via the peripheral nervous system. The CNS integrates this information and coordinates output commands to maintain energy balance. As mentioned above, hormonal feedback helps inform the CNS about peripheral energy stores and energy availability. Ghrelin is one such metabolic hormone that signals the brain to control energy balance.

Ghrelin is a 28 amino acid peptide predominantly synthesized in the stomach, where it is secreted into the circulation. It is a potent stimulator of growth hormone release and enhances feeding and weight gain to regulate energy homeostasis. The growth hormone secretagogue receptor (GHSR1a) is the key receptor through which ghrelin mediates these effects. Pro-ghrelin requires posttranslational acylation with *n*-octanoic acid or n-decanoic acid at the third serine for its biological activity at the GHSR1a. Thus, ghrelin exists as two forms in the plasma, acylated ghrelin and des-acylated ghrelin. Ghrelin O-acyltransferase (GOAT) is the enzyme responsible for pro-ghrelin acylation (Yang et al. 2008) and is also found predominantly in the stomach and digestive tract (Gutierrez et al. 2008; Yang et al. 2008). In the stomach and duodenum GOAT co-localizes with ghrelin expressing cells (Sakata et al. 2009), where it can readily acylate newly synthesized pro-ghrelin. GOAT can acylate pro-ghrelin with other fatty acid substrates besides octanoate and this is likely a function of dietary fatty acid availability (Kirchner et al. 2009). Once ghrelin is acetylated, it is transported to the Golgi apparatus and cleaved by prohormone convertase 1/3 (PC 1/3) to form 28 amino acid mature ghrelin (Zhu et al. 2006). Although des-acyl ghrelin is at high concentrations in the plasma it does not activate GHSR1a. The GHSR1a is the only functional ghrelin receptor that has been effectively charac-terized. It is a G protein-coupled 7-transmembrane receptor and is required to elicit growth hormone release or a food intake response to exogenous administered ghrelin. This chapter focuses on the intracellular signaling mechanisms the GHSR1a utilizes in the hypothalamus to regulate energy balance.

GHSR1a Expressing Nuclei in the Hypothalamus

The GHSR1a is expressed in many hypothalamic nuclei with the highest expression in the arcuate nucleus (ARC) (Zigman et al. 2006). The ARC plays an important role in appetite regulation and body weight and there are two key appetite-regulating neuronal populations in ARC (see Fig. 1). Neuropeptide Y (NPY) and agouti-related peptide AgRP are co-expressed in ARC neurons and are potent orexigenic peptides, whereas the proopiomelanocortin (POMC) precursor protein is cleaved into the potent anorexigenic α-melanocyte-stimulating hormone (α-MSH) peptide. AgRP and POMC neurons in the ARC are arguably considered "first-order" sensory neurons in the control of food intake and receive, coordinate, and respond to changes in metabolic status. Both AgRP and POMC neurons project to the PVN, where the anorectic effects of α-MSH peptides are mediated by melanocortin 4 receptors (MC4R). NPY Y1 and Y5 receptors in the PVN mediate the orexigenic effects of NPY, whereas AgRP antagonizes the effect of α-MSH on the MC4R. A unique feature of the melanocortin system is the ability of AgRP neurons to suppress POMC cell firing via inhibitory GABAergic inputs (Andrews et al. 2008; Cowley et al. 2003). There is no evidence that POMC neurons feed back to inhibit AgRP neuronal firing despite the expression of GABA in POMC neurons (Hentges et al. 2004, 2009; Aponte et al. 2011; Atasoy et al. 2012). The fact that the GHSR1a is expressed on >94 % of orexigenic AgRP neurons and <8 % of anorectic POMC (Willesen et al. 1999) underlies the appetite-stimulating effects of ghrelin. Moreover, the GHSR1a is expressed on approximately 25 % of somatostatin and 30 % of growth hormone releasing hormone neurons in the ARC (Willesen et al. 1999). This partially regulates the actions of ghrelin on growth hormone release from the pituitary although recent studies from our lab show that >80 % of pituitary somatotrophs express the GHSR1a (Reichenbach et al. 2012).

There is also moderate expression of the GHSR1a in the PVN (Guan et al. 1997; Zigman et al. 2006), another important nucleus controlling appetite and adiposity. The PVN receives projections from numerous hypothalamic, limbic, and cortical nuclei, including NPY, AgRP, and POMC neurons in the ARC. The PVN also has strong efferent outputs to peripheral organs via the autonomic nervous system, thus it is possible that GHSR1a mRNA expression in the PVN fine-tunes the ARC NPY and AgRP inputs after receiving information from other hypothalamic or higher limbic and cortical regions. In this way, other physiological cues, such as emotionality and stress, can modulate appetite and energy status by affecting PVN outputs. Indeed, ghrelin is known to strongly influence the stress axis (Spencer et al. 2012) and regulates feeding behavior under chronically stressed conditions (Chuang et al. 2010, 2011).

A previous study demonstrated that knockdown of *GHSR* mRNA in the PVN reduces body weight and blood ghrelin levels without affecting food intake (Shrestha et al. 2009). This suggests a divergence between energy intake and body weight and that ghrelin acts via NPY release in the paraventricular nucleus to promote food intake, but that ghrelin works directly in the paraventricular nucleus to promote adiposity. However, it should be noted that direct injection of ghrelin into the PVN increases food intake (Olszewski et al. 2003)

The dorsomedial hypothalamic nucleus (DMH) also expresses moderate levels of GHSR1a mRNA although the role of the GHSR in the DMH is almost completely unknown. One recent study using GHSR1a knockout mice examined cfos activation in the hypothalamus during a scheduled feeding paradigm. These results

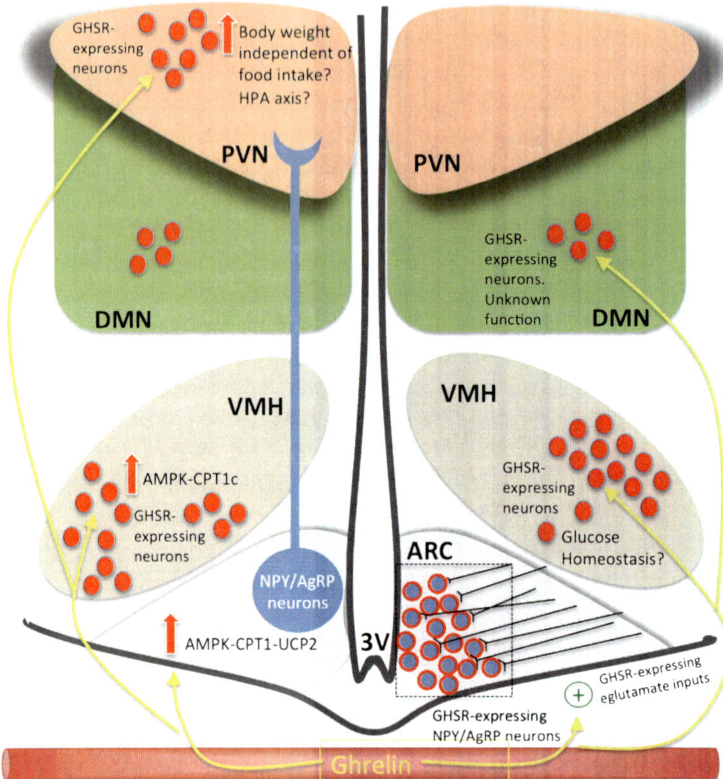

Fig. 1 Ghrelin targets ghrelin receptor (GHSR) expressing neurons in the hypothalamus. Ghrelin acts on GHSRs in the ARC to increase appetite via two mechanisms. First, ghrelin increases a AMPK-CPT1-UCP2 pathway in NPY/AgRP neurons and second, ghrelin increases glutamatergic inputs onto NPY/AgRP neurons. Ghrelin also acts on GHSRs in the VMH to increase appetite via an AMPK-dependent mechanism. The role of GHSRs in the DMN and PVN remain unknown although studies suggest that GHSRs in PVN increase body weight independent from food intake

showed that GHSR1a knockout mice exhibited reduced cfos activation in the DMH compared to controls suggesting that GHSR1a in the DMH may mediate anticipation of a meal (Blum et al. 2009). However, it should be noted that the same cfos pattern between controls and knockouts was observed in the ARC, PVN, and lateral hypothalamus, and cfos activation can occur via neural inputs from different nuclei and does not guarantee direct activation via hormonal inputs. As yet, the role of GHSR1a in the DMH remains to be determined and recent techniques in molecular genetics may elucidate the role of the receptor in the DMH. One potential unexplored function may be the activation of neurons that release gonadotropin-inhibitory hormone, as these neurons inhibit reproduction and stimulate feeding (Clarke et al. 2012). Similarly, ghrelin suppresses the reproductive axis and increases feeding (Smith et al. 2013; Furuta et al. 2001; Kluge et al. 2012; Ogata et al. 2009).

The ventromedial hypothalamic nucleus (VMH) is a brain region known to regulate appetite, body weight, and glucose homeostasis. For example, studies show that the VMH provides a strong excitatory input to POMC neurons and fasting diminishes the strength of the excitatory input from the VMH to POMC neurons (Sternson et al. 2005). Deletion of leptin receptors from POMC neurons (Balthasar et al. 2004) or re-expression of leptin receptors on POMC neurons shows a minor effect on food intake (Huo et al. 2009), but a strong effect on whole body glucose metabolism (Huo et al. 2009). The effect of leptin to inhibit food intake may be driven largely through the VMH input onto POMC neurons as deleting glutamate synaptic transmission from VMH neurons increased long-term food intake and susceptibility to diet-induced obesity (Tong et al. 2007). While the studies discussed above show that the VMH has strong inhibitory influence over appetite, the VMH expresses moderate levels of GHSR1a and ghrelin injection in the VMH increases food intake (Lopez et al. 2008).

The lateral hypothalamus (LH) was labeled as the "feeding centre" when Anand and Brobeck showed that bilateral destruction of the LH completely suppressed spontaneous eating (Anand and Brobeck 1951). More recently, studies show that orexin neurons in the LH regulate the midbrain dopamine reward (Borgland et al. 2006) including food rewards (DiLeone et al. 2012). Moreover, ghrelin increases food intake by modulating the reward-related motivation in the midbrain dopamine system (Abizaid 2009; Abizaid et al. 2006; Naleid et al. 2005). While ghrelin knockout mice show reduced orexin neurons (Lamont et al. 2012) and ghrelin modulates the reward value of high fat food partially via orexin receptors (Perello et al. 2010), there is no good evidence to support the presence of GHSR1a in the LH (Guan et al. 1997; Zigman et al. 2006). This suggests that ghrelin sensitive areas, such as the ARC, transmit ghrelin-related signals via neural inputs into the LH. One possibility is NPY and AgRP neurons in the ARC, as both of which have terminal projections in the LH (Dube et al. 1999; Horvath et al. 1999; Toshinai et al. 2003).

Ghrelin Activates Hypothalamic Circuits that Control Food Intake

Ghrelin induces feeding by robustly stimulating NPY and AgRP neuronal activity as assessed by electrophysiology (Andrews et al. 2008; Cowley et al. 2003) or fos immunoreactivity (Andrews et al. 2008; Hewson and Dickson 2000; Wang et al. 2002) and gene expression (Chen et al. 2004; Kamegai et al. 2000, 2001; Nakazato et al. 2001). Genetic ablation of AgRP in adulthood abolishes the orexigenic effects of ghrelin (Luquet et al. 2007) and double NPY/AgRP knockout mice do not increase food intake in response ghrelin (Chen et al. 2004).

At the same time that ghrelin stimulates orexigenic NPY/AgRP neuronal activity, POMC neuronal activity is suppressed via inhibitory γ-aminobutyric acid (GABA)-eric inputs from active NPY/AgRP neurons (Cowley et al. 2003).

Deletion of vesicular GABA transporter in AgRP neurons removes the inhibitory tone onto postsynaptic POMC cells, allowing unopposed activation of the melanocortin system and subsequent anorexia (Tong et al. 2008). GABA-mediated electrophysiological inhibition of POMC neurons by NPY/AgRP neurons is accompanied by changes in POMC neuronal synaptic plasticity, in which ghrelin increases the number of inhibitory perikaryal synapses on POMC neurons (Andrews et al. 2008). Increased GABAergic inhibitory inputs on POMC neurons favors elevated food intake by lowering anorexigenic POMC neuronal activity.

How does Ghrelin Activate Appetite-Stimulating Neurons?

Recent evidence has begun to unravel how ghrelin activates NPY neurons to initiate changes in feeding behavior. The unique intracellular signaling modality connects mitochondrial-mediated effects of G-coupled receptors on neuronal function and associated feeding behavior (see Fig. 2).

AMPK

AMPK was identified as a critical component of the signaling mode. AMPK is an intracellular energy sensor that switches off ATP-consuming pathways and switches on ATP-producing pathways such as glucose uptake and fatty acid oxidation (Steinberg and Kemp 2009). AMPK is a heterotrimeric complex of three enzymatic subunits (α, β, γ) that can be activated by upstream kinases such as the tumor suppressor, LKB1, or calmodulin (CaM)-dependent protein kinase kinases (CaMKK) (Steinberg and Kemp 2009). LKB1 activation of AMPK is dependent upon 5′-AMP binding to the γ-subunit, whereas CaMKK-dependent activation of AMPK is independent of AMP and requires an increase in intracellular Ca^{2+}.

Both intraperitoneal or intracerebroventricular ghrelin injection increased AMPK phosphorylation and activity in the hypothalamus (Andersson et al. 2004; Kola et al. 2005; Andrews et al. 2008) and increased food intake. Inhibition of AMPK activity with compound C reduced ghrelin stimulated food intake and ghrelin does not activate AMPK in Ghsr-/- mice (Andrews et al. 2008; Lopez et al. 2008). These results illustrate that ghrelin acts upstream of AMPK to stimulate food intake. While the studies above highlight ghrelin's stimulatory effect on AMPK and food intake in hypothalamic extracts and live animals, ghrelin also activates AMPK in isolated NPY neurons (Kohno et al. 2008), which adds the desired anatomical specificity to the NPY/AgRP circuitry controlling food intake. In addition, ghrelin regulates food intake by increasing AMPK activity in the VMH, independent of AMPK activity in the ARC (Lopez et al. 2008).

Furthermore, increasing AMPK activity in the mediobasal hypothalamus, using a constitutively active adenoviral approach, increased body weight and food intake (Minokoshi et al. 2004). The increase in food intake was associated with increased

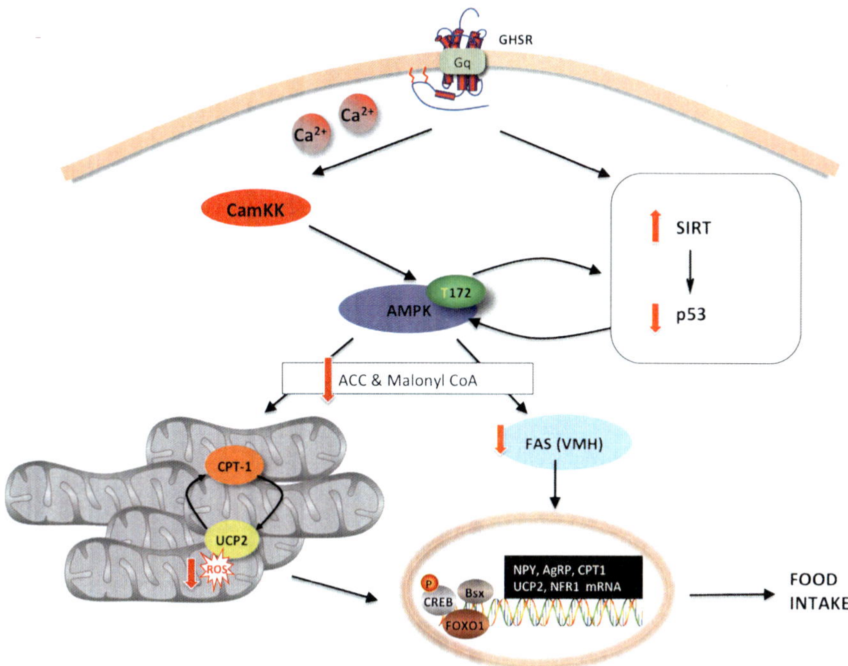

Fig. 2 Activation of the ghrelin receptor (GHSR) increases AMPK activity either via CamKK or SIRT1. This leads to gene expression of peptides known to increase appetite such as NPY and AgRP

NPY and AgRP gene expression under fasted conditions (Minokoshi et al. 2004). Selective deletion of the AMPK catalytic alpha subunit (AMPK α2) in AgRP neurons decreased body weight gain and showed greater sensitivity to suppress food intake after injection of a melanocortin agonist (Claret et al. 2007). Collectively, these results indicate that AMPK in NPY/AgRP neurons is an important regulator of food intake and body weight and that ghrelin uses AMPK to stimulate food intake. Moreover, ghrelin-induced AMPK activity decreases the mammalian target of rapamycin 1 (mTORC1) activity and increases AgRP mRNA (Watterson et al. 2013), consistent with the effect of hypothalamic mTOR on appetite (Cota et al. 2006).

Calcium and CaM-Dependent Protein Kinase Kinase

In order to activate AMPK activity, ghrelin binds to the GHSR and initiates Ca^{2+} influx and/or release in identified NPY neurons (Kohno et al. 2003, 2007, 2008). Ghrelin can increase intracellular Ca^{2+} through multiple signaling pathways including adenylate cyclase, cAMP, phospholipase C (Kohno et al. 2007), protein kinase A, and N-type Ca^{2+} channels (Kohno et al. 2003). The increase in intracellular Ca^{2+} interacts with calmodulin (CaM) to activate CaM-dependent protein kinase kinases (CaMKK). Because CaMKK is an upstream kinase that can activate AMPK (Woods et al. 2005), Anderson et al. used CamKK-/- mice and showed that ghrelin did not stimulate feeding (Anderson et al. 2008). Moreover, the results showed that a rise in intracellular Ca^{2+} caused CaMKK activation, which in turn led to AMPK phosphorylation and increased NPY protein and message (Anderson et al. 2008). Taken together, these studies suggest that ghrelin increases intracellular Ca^{2+}, leading to activation of CaMKK and subsequent AMPK phosphorylation. However, other studies show that ghrelin activates AMPK in NPY neurons, which then leads to subsequent Ca^{2+} release (Kohno et al. 2008). In this study, CaMKK cannot be the upstream kinase responsible for AMPK activation as intracellular Ca^{2+} rises after AMPK activation. The authors speculative that the LKB1 may be the upstream kinase, as cAMP-PKA signaling leads to activation of LKB1 (Collins et al. 2000; Yin et al. 2003) and ghrelin increases intracellular Ca^{2+} in isolated NPY neurons through cAMP and PKA-dependent mechanisms (Kohno et al. 2003). Clearly, more evidence is required to clarify this issue, however, the GHSR receptor is a G-coupled receptor that interacts with $G\alpha q$, which in turn recruits phospholipase C ($PLC\beta$) to the membrane and catalyzes the enzymatic production of inositol-1,4,5-triphosphate (IP3). As IP3 is a soluble second messenger that initiates Ca^{2+} release from the endoplasmic reticulum, it seems likely that CaMKK is the critical upstream kinase required for AMPK activation in NPY neurons. One particular caveat must be mentioned here, all of the work on GHSR signaling via a $G\alpha q$-$PLC\beta$-IP3 pathway has been characterized in either transfected cell culture models using often COS or HEK cells or in anterior pituitary somatotrophs. Whether or not neurons employ the same pathway remains to be determined, however, an identical mechanism is likely based on the importance of Ca^{2+} to stimulate CaMKK (Anderson et al. 2008), AMPK, and downstream pathways described below.

SIRT1

Velasquez et al. recently showed that SIRT1 deacetylase activity in the hypothalamus also mediates ghrelin-induced AMPK activity and food intake (Velasquez et al. 2011). SIRT1 is an NAD+ dependent deacetylase that acts to deacetylate p53, and p53 is hyperacetylated in SIRT1 knockout mice limiting its function (Han et al. 2008). p53 is relevant in the context of feeding as increased AMPK activity activates p53 (Vousden and Ryan 2009). Both fasting and ghrelin increase SIRT1 deacetylase activity in the hypothalamus (Velasquez et al. 2011) similar to the regulation of SIRT1 activity by nutritional status in peripheral tissues (Cohen et al. 2004). Velasquez et al. showed that a SIRT1 inhibitor reduced ghrelin-induced pAMPK, NPY, and AgRP mRNA (Velasquez et al. 2011). Further, they used p53 knockout mice, as p53 is a substrate of SIRT1, and showed that ghrelin does not increase food intake or AMPK activity (Velasquez et al. 2011). However, AICAR still increased food intake in p53 knockout mice (Velasquez et al. 2011) indicating that SIRT1 acts upstream of AMPK. It is unknown if

CaMKK and SIRT1 interact upstream of AMPK to regulate AMPK activation. SIRT1 deletion in AgRP neurons also reduced action potential firing in response to ghrelin further supporting and key role for SIRT1 in ghrelin-induced AgRP neuronal function (Dietrich et al. 2010).

Carnitine Palmitoyl Transferase 1

The downstream intracellular actions after ghrelin-induced AMPK activation involve phosphorylation of acetyl CoA carboxylase (ACC), which causes the suppression of malonyl CoA and disinhibition of carnitine palmitoyl transferase 1 (CPT1). Increased CPT1 increases fatty acid Acyl-CoA transport into mitochondria for oxidation, whereas malonyl CoA acts as an allosteric inhibitor of CPT1 and thus prevents fatty acid Acyl-CoA transport into mitochondria for oxidation. Several studies indicate that the enzymes involved in fat metabolism play an important role in the hypothalamic regulation of food intake. Increased hypothalamic malonyl CoA reduces food intake and is regulated acutely by leptin and glucose (Wolfgang et al. 2007). Moreover, lowering hypothalamic malonyl CoA with either ACC inhibitors or viral overexpression of malonyl CoA decarboxylase (MCD), increases food intake and body weight gain and malonyl CoA is reduced during fasting and elevated after feeding indicating it plays a physiological role in food intake (Wolfgang and Lane 2006). In 2003, Obici et al. (Obici et al. 2003) reported that inhibition of central CPT1a, either by pharmacological or genetic knockdown, inhibited food intake and peripheral glucose production. The brain-specific CPT1c isoform also regulates peripheral energy metabolism, as CPT1c-/- mice exhibit reduced body weight gain, decreased food intake, and glucose intolerance (Gao et al. 2009; Wolfgang et al. 2006). Because ghrelin activates AMPK, which is upstream of CPT1, it was hypothesized that CPT1 mediates ghrelin-induced food intake (Andrews et al. 2008; Lopez et al. 2008) through a malonyl CoA-dependent mechanism. Ghrelin suppressed malonyl CoA and increased CPT1 activity and protein in the hypothalamus after 2 h but not 6 h and pharmacological inhibition of CPT1 prevented ghrelin-induced food intake (Lopez et al. 2008). Inhibition of CPT1 also prevented ghrelin's ability to increase NPY and AgRP mRNA expression in the hypothalamus (Andrews et al. 2008).

Recent studies show that ghrelin does not increase food intake CPT1c-/- mice despite high levels of AMPK activity (Ramirez et al. 2013), again highlighting that CPT1c is downstream of AMPK. These authors demonstrated that ghrelin surprisingly increased ceramide synthesis via a CPT1c-dependent process. Notably, the orexigenic effects of ghrelin were blocked by central inhibition of ceramide synthesis using the inhibitor myriocin. Myriocin also blocked the ghrelin-induced rise in NPY and AgRP, as well as key transcription factors in the ARC FOXO1 and cAMP-response element binding protein. CPT1c and ceramide synthesis may be a common target for hormonal regulation of feeding neurons as leptin reduced hypothalamic ceramide. Moreover, overexpression of CPT1c in the hypothalamus prevented the ability of leptin to suppress food intake and NPY mRNA levels (Gao et al. 2011).

Uncoupling Protein 2

CPT1 transports fatty acid Acyl-CoA to mitochondria for oxidation, therefore ghrelin-induced activation of the AMPK-CPT1 axis should lead to subsequent changes in mitochondrial respiration. Indeed, ghrelin stimulated palmitate-driven uncoupled respiration in isolated hypothalamic mitochondria (Andrews et al. 2008) in an UCP2-dependent fashion, as no effect was observed in UCP2-/- mice. Moreover, AMPK is required to activate this UCP2-dependent mitochondrial mechanism and UCP2 is required to permit CPT1 activation. Thus, upon binding to its receptor, ghrelin activates this AMPK-CPT1-UCP2 axis and initiates a mitochondrial mechanism that is essential for mitochondrial biogenesis in AgRP neurons, electrical activation of AgRP neurons, and for ghrelin-triggered synaptic plasticity of POMC. Collectively, this causes ghrelin-induced food intake. It is worthwhile noting that UCP2 decreases the ATP generating potential in pancreatic beta cells resulting in reduced glucose-stimulated insulin release (Zhang et al. 2001), raising the possibility that activation of UCP2 in NPY neurons also suppresses the ATP generating potential of an individual mitochondrion. There is currently no experimental evidence regarding ATP production in NPY neurons after UCP2 activation, however, we believe the increase in mitochondrial biogenesis would overcome any decrease in ATP generating potential per mitochondrion. Indeed, the ability of ghrelin to increase mitochondrial biogenesis in a UCP2-dependent manner in NPY or dopamine neurons supports this hypothesis (Andrews et al. 2008, 2009).

Furthermore, ghrelin not only increases the AMPK-CPT1-UCP2 fatty acid oxidation pathway, it also initially increases fatty acyl-CoA concentration in the hypothalamus, as a substrate for fatty acid oxidation in mitochondria (Andrews et al. 2008). Reactive oxygen species (ROS) are a byproduct of enhanced mitochondrial respiration during fat acid oxidation and UCP2 is a mitochondrial protein primarily known to buffer or scavenge excessive ROS production (Andrews et al. 2005). It was discovered that UCP2, specifically in NPY/AgRP neurons, is required to buffer excessive ROS production generated by ghrelin-induced fatty acid oxidation (Andrews et al. 2008). Thus, ghrelin activation of this AMPK-CPT1-UCP2 pathway permits increased fatty acid oxidation while buffering increased ROS in NPY neurons, but not POMC neurons because they lack GHSRs. This increase in mitochondrial activity and ROS buffering permits an increase in NPY and AgRP mRNA gene expression and the bioenergetic capacity to sustain NPY/AgRP cell firing and maintain a hunger signal during periods of negative energy balance and starvation. This appears to be a selective advantage to maintain NPY/AgRP cell function especially considering that ablation of NPY/AgRP results in starvation and death, and ablation of POMC "only" results in obesity (Gropp et al. 2005; Luquet et al. 2005).

Synaptic Regulation of Feeding Circuits: A Role for Ghrelin and AMPK

Recent evidence shows that synaptic plasticity within the hypothalamus mediates appetite and body weight under varying metabolic states (Horvath 2005). For example, ob/ob mice had increased excitatory synapses and decreased inhibitory synapses on AgRP neurons, whereas POMC neurons showed reduced excitatory synapses. This arrangement favors AgRP activation and subsequent hyperphagia. Leptin treatment to ob/ob mice normalized the synaptic input organization to wild-type levels within 6 h (Pinto et al. 2004). On the other hand, ghrelin shifted the synaptic profile of POMC neurons in the opposite direction caused by leptin. Ghrelin decreased inhibitory inputs on POMC neurons thereby reducing satiety drive through reduced activation of POMC neurons (Andrews et al. 2008). Although hormonal feedback mechanisms influence synaptic plasticity in AgRP or POMC neurons, the intracellular mechanisms that induce rapid and persistent changes in synaptic activity and connectivity remained enigmatic until Yang et al. (Yang et al. 2011) recently provided the first mechanistic clues. Initially, the authors showed food deprivation increases action potential firing frequency in AgRP neurons, which was dependent on glutamatergic excitatory inputs. Intriguingly, AMPK activity stimulates internal calcium release via the ryanodine receptor and underpins the excitatory activity in presynaptic nerve terminals. Ghrelin also triggered an AMPK-dependent positive feedback loop in presynaptic terminals that sustained excitatory synaptic activity hours after ghrelin removal. Leptin switched off this persistent activity by activating an opioid receptor-dependent mechanism in which POMC neurons release opioid peptides.

Because synaptic plasticity induces a potential memory capacity (Gordon and Bains 2006), Yang et al. postulate their observations confer a presynaptic memory storage system that regulates AgRP neuronal firing under different metabolic states and hormone exposure. Ghrelin functions optimally under conditions of negative energy balance (Briggs and Andrews 2011) to shift an organism toward neutral energy balance. This ghrelin-induced excitatory synaptic activity to AgRP neurons would therefore be an essential mechanism to ensure continuous AgRP cell firing under negative energy balance. The synaptic memory capacity also ensures that AgRP neurons continue to fire as ghrelin levels recede. This is a critical point as high ghrelin levels from food restriction fall within 2 h of refeeding (Tschop et al. 2000) and achieving neutral energy homeostasis may take a longer period of time.

These data conceptually support studies showing that genetic deletion of AgRP neurons in adulthood results in starvation and death (Gropp et al. 2005; Luquet et al. 2005). From an evolutionary standpoint, this synaptic AMPK-dependent positive feedback loop maintains AgRP firing and a hunger stimulus during periods of food scarcity and promotes food intake to ensure survival. Given that periods of negative energy balance dominated evolutionary history, it is not surprising that AgRP neurons developed different molecular mechanisms, compared to POMC neurons, to preserve cell function and appetitive drive.

There are a few important caveats to this study. First, all experiments were conducted in vitro and under these conditions the experimental environment is drastically different from the physiological condition in situ. For example, in these in vitro studies glucose was clamped at 11 mM, which is significantly higher than either the fed or fasted state in situ. Indeed, varying levels of glucose have differential effects on AgRP neuronal firing in mice lacking AMPK activity in AgRP neurons (Claret et al. 2007). Second, the identity of the cells providing the presynaptic input is unknown. While these presynaptic cells must contain the GHSR, it is unknown if ghrelin acts on the cell body or locally on presynaptic terminals expressing the GHSR. Finally, this presynaptic mechanism must operate in a synergistic manner with a direct hormonal effect on AgRP neurons, as ghrelin increases AMPK activity in isolated AgRP neurons (Kohno et al. 2008). Moreover, leptin suppresses ghrelin-induced firing of isolated AgRP neurons (Kohno et al. 2007) and declining leptin or glucose concentrations activates AMPK activity in GFP-identified NPY neurons (Murphy et al. 2009).

Future Directions

Although ghrelin was discovered in 1999, there still remains a lot to be determined about GHSR1a function in the hypothalamus. There are no reliable antibodies to stain GHSR1a in the hypothalamus and as a consequence, we know very little about the hypothalamic neuronal populations expressing the receptor. The development of novel genetic models and tools will be vital to further our understanding about the neuroanatomy of the GHSR1a circuits in the hypothalamus. Pharmacogenetic and optogenetic tools will also help shed light on the function of the GHSR1a in hypothalamic nuclei such as the VMH and DMH.

Acknowledgments This work was supported by a Monash Fellowship, Monash University, Australia, an Australia Research Council Future Fellowship (FT 100100966) and NHMRC grants (NHMRC 1011274, NHMRC 1030037) to ZBA.

References

Abizaid A (2009) Ghrelin and dopamine: new insights on the peripheral regulation of appetite. J Neuroendocrinol 21(9):787–793

Abizaid A, Liu ZW, Andrews ZB, Shanabrough M, Borok E, Elsworth JD, Roth RH, Sleeman MW, Picciotto MR, Tschop MH, Gao XB, Horvath TL (2006) Ghrelin modulates the activity and synaptic input organization of midbrain dopamine neurons while promoting appetite. J Clin Invest 116(12):3229–3239

Anand BK, Brobeck JR (1951) Hypothalamic control of food intake in rats and cats. Yale J Biol Med 24(2):123–140

Anderson KA, Ribar TJ, Lin F, Noeldner PK, Green MF, Muehlbauer MJ, Witters LA, Kemp BE, Means AR (2008) Hypothalamic CaMKK2 contributes to the regulation of energy balance. Cell Metab 7(5):377–388

Andersson U, Filipsson K, Abbott CR, Woods A, Smith K, Bloom SR, Carling D, Small CJ (2004) AMP-activated protein kinase plays a role in the control of food intake. J Biol Chem 279(13):12005–12008

Andrews ZB, Diano S, Horvath TL (2005) Mitochondrial uncoupling proteins in the CNS: in support of function and survival. Nat Rev Neurosci 6(11):829–840

Andrews ZB, Erion D, Beiler R, Liu ZW, Abizaid A, Zigman J, Elsworth JD, Savitt JM, DiMarchi R, Tschoep M, Roth RH, Gao XB, Horvath TL (2009) Ghrelin promotes and protects nigrostriatal dopamine function via a UCP2-dependent mitochondrial mechanism. J Neurosci 29(45):14057–14065

Andrews ZB, Liu ZW, Walllingford N, Erion DM, Borok E, Friedman JM, Tschop MH, Shanabrough M, Cline G, Shulman GI, Coppola A, Gao XB, Horvath TL, Diano S (2008) UCP2 mediates ghrelin's action on NPY/AgRP neurons by lowering free radicals. Nature 454(7206):846–851

Aponte Y, Atasoy D, Sternson SM (2011) AGRP neurons are sufficient to orchestrate feeding behavior rapidly and without training. Nat Neurosci 14(3):351–355

Atasoy D, Betley JN, Su HH, Sternson SM (2012) Deconstruction of a neural circuit for hunger. Nature 488(7410):172–177

Balthasar N, Coppari R, McMinn J, Liu SM, Lee CE, Tang V, Kenny CD, McGovern RA, Chua SC Jr, Elmquist JK, Lowell BB (2004) Leptin receptor signaling in POMC neurons is required for normal body weight homeostasis. Neuron 42(6):983–991

Blum ID, Patterson Z, Khazall R, Lamont EW, Sleeman MW, Horvath TL, Abizaid A (2009) Reduced anticipatory locomotor responses to scheduled meals in ghrelin receptor deficient mice. Neuroscience 164(2):351–359

Borgland SL, Taha SA, Sarti F, Fields HL, Bonci A (2006) Orexin A in the VTA is critical for the induction of synaptic plasticity and behavioral sensitization to cocaine. Neuron 49(4):589–601

Briggs DI, Andrews ZB (2011) Metabolic status regulates ghrelin function on energy homeostasis. Neuroendocrinology 93(1):48–57

Chen HY, Trumbauer ME, Chen AS, Weingarth DT, Adams JR, Frazier EG, Shen Z, Marsh DJ, Feighner SD, Guan XM, Ye Z, Nargund RP, Smith RG, Van der Ploeg LH, Howard AD, MacNeil DJ, Qian S (2004) Orexigenic action of peripheral ghrelin is mediated by neuropeptide Y and agouti-related protein. Endocrinology 145(6):2607–2612

Chuang JC, Cui H, Mason BL, Mahgoub M, Bookout AL, Yu HG, Perello M, Elmquist JK, Repa JJ, Zigman JM, Lutter M (2010) Chronic social defeat stress disrupts regulation of lipid synthesis. J Lipid Res 51(6):1344–1353

Chuang JC, Perello M, Sakata I, Osborne-Lawrence S, Savitt JM, Lutter M, Zigman JM (2011) Ghrelin mediates stress-induced food-reward behavior in mice. J Clin Investig 121(7):2684–2692

Claret M, Smith MA, Batterham RL, Selman C, Choudhury AI, Fryer LG, Clements M, Al-Qassab H, Heffron H, Xu AW, Speakman JR, Barsh GS, Viollet B, Vaulont S, Ashford ML, Carling D, Withers DJ (2007) AMPK is essential for energy homeostasis regulation and glucose sensing by POMC and AgRP neurons. J Clin Invest 117(8):2325–2336

Clarke IJ, Smith JT, Henry BA, Oldfield BJ, Stefanidis A, Millar RP, Sari IP, Chng K, Fabre-Nys C, Caraty A, Ang BT, Chan L, Fraley GS (2012) Gonadotropin-inhibitory hormone is a hypothalamic peptide that provides a molecular switch between reproduction and feeding. Neuroendocrinology 95(4):305–316

Cohen HY, Miller C, Bitterman KJ, Wall NR, Hekking B, Kessler B, Howitz KT, Gorospe M, de Cabo R, Sinclair DA (2004) Calorie restriction promotes mammalian cell survival by inducing the SIRT1 deacetylase. Science 305(5682):390–392

Collins SP, Reoma JL, Gamm DM, Uhler MD (2000) LKB1, a novel serine/threonine protein kinase and potential tumour suppressor, is phosphorylated by cAMP-dependent protein kinase (PKA) and prenylated in vivo. Biochem J 345(Pt 3):673–680

Cota D, Proulx K, Smith KA, Kozma SC, Thomas G, Woods SC, Seeley RJ (2006) Hypothalamic mTOR signaling regulates food intake. Science 312(5775):927–930

Cowley MA, Smith RG, Diano S, Tschop M, Pronchuk N, Grove KL, Strasburger CJ, Bidlingmaier M, Esterman M, Heiman ML, Garcia-Segura LM, Nillni EA, Mendez P, Low MJ, Sotonyi P, Friedman JM, Liu H, Pinto S, Colmers WF, Cone RD, Horvath TL (2003) The distribution and mechanism of action of ghrelin in the CNS demonstrates a novel hypothalamic circuit regulating energy homeostasis. Neuron 37(4):649–661

Dietrich MO, Antunes C, Geliang G, Liu ZW, Borok E, Nie Y, Xu AW, Souza DO, Gao Q, Diano S, Gao XB, Horvath TL (2010) Agrp neurons mediate Sirt1's action on the melanocortin system and energy balance: roles for Sirt1 in neuronal firing and synaptic plasticity. J Neurosci 30(35):11815–11825

DiLeone RJ, Taylor JR, Picciotto MR (2012) The drive to eat: comparisons and distinctions between mechanisms of food reward and drug addiction. Nat Neurosci 15(10):1330–1335

Dube MG, Kalra SP, Kalra PS (1999) Food intake elicited by central administration of orexins/ hypocretins: identification of hypothalamic sites of action. Brain Res 842(2):473–477

Farooqi S, O'Rahilly S (2006) Genetics of obesity in humans. Endocr Rev 27(7):710–718

Furuta M, Funabashi T, Kimura F (2001) Intracerebroventricular administration of ghrelin rapidly suppresses pulsatile luteinizing hormone secretion in ovariectomized rats. Biochem Biophys Res Commun 288(4):780–785

Gao S, Zhu G, Gao X, Wu D, Carrasco P, Casals N, Hegardt FG, Moran TH, Lopaschuk GD (2011) Important roles of brain-specific carnitine palmitoyltransferase and ceramide metabolism in leptin hypothalamic control of feeding. Proc Natl Acad Sci USA 108(23):9691–9696

Gao XF, Chen W, Kong XP, Xu AM, Wang ZG, Sweeney G, Wu D (2009) Enhanced susceptibility of Cpt1c knockout mice to glucose intolerance induced by a high-fat diet involves elevated hepatic gluconeogenesis and decreased skeletal muscle glucose uptake. Diabetologia 52(5):912–920

Gordon GR, Bains JS (2006) Can homeostatic circuits learn and remember? J Physiol 576(Pt 2):341–347

Gropp E, Shanabrough M, Borok E, Xu AW, Janoschek R, Buch T, Plum L, Balthasar N, Hampel B, Waisman A, Barsh GS, Horvath TL, Bruning JC (2005) Agouti-related peptide-expressing neurons are mandatory for feeding. Nat Neurosci 8(10):1289–1291

Guan XM, Yu H, Palyha OC, McKee KK, Feighner SD, Sirinathsinghji DJ, Smith RG, Van der Ploeg LH, Howard AD (1997) Distribution of mRNA encoding the growth hormone secretagogue receptor in brain and peripheral tissues. Brain Res Mol Brain Res 48(1):23–29

Gutierrez JA, Solenberg PJ, Perkins DR, Willency JA, Knierman MD, Jin Z, Witcher DR, Luo S, Onyia JE, Hale JE (2008) Ghrelin octanoylation mediated by an orphan lipid transferase. Proc Natl Acad Sci USA 105(17):6320–6325

Han MK, Song EK, Guo Y, Ou X, Mantel C, Broxmeyer HE (2008) SIRT1 regulates apoptosis and Nanog expression in mouse embryonic stem cells by controlling p53 subcellular localization. Cell Stem Cell 2(3):241–251

Hentges ST, Nishiyama M, Overstreet LS, Stenzel-Poore M, Williams JT, Low MJ (2004) GABA release from proopiomelanocortin neurons. J Neurosci: Off J Soc Neurosci 24(7):1578–1583

Hentges ST, Otero-Corchon V, Pennock RL, King CM, Low MJ (2009) Proopiomelanocortin expression in both GABA and glutamate neurons. J Neurosci: Off J Soc Neurosci 29(43):13684–13690

Hewson AK, Dickson SL (2000) Systemic administration of ghrelin induces Fos and Egr-1 proteins in the hypothalamic arcuate nucleus of fasted and fed rats. J Neuroendocrinol 12(11):1047–1049

Horvath TL (2005) The hardship of obesity: a soft-wired hypothalamus. Nat Neurosci 8(5):561–565

Horvath TL, Diano S, van den Pol AN (1999) Synaptic interaction between hypocretin (orexin) and neuropeptide Y cells in the rodent and primate hypothalamus: a novel circuit implicated in metabolic and endocrine regulations. J Neurosci 19(3):1072–1087

Huo L, Gamber K, Greeley S, Silva J, Huntoon N, Leng XH, Bjorbaek C (2009) Leptin-dependent control of glucose balance and locomotor activity by POMC neurons. Cell Metab 9(6):537–547

Kamegai J, Tamura H, Shimizu T, Ishii S, Sugihara H, Wakabayashi I (2000) Central effect of ghrelin, an endogenous growth hormone secretagogue, on hypothalamic peptide gene expression. Endocrinology 141(12):4797–4800

Kamegai J, Tamura H, Shimizu T, Ishii S, Sugihara H, Wakabayashi I (2001) Chronic central infusion of ghrelin increases hypothalamic neuropeptide Y and Agouti-related protein mRNA levels and body weight in rats. Diabetes 50(11):2438–2443

Kirchner H, Gutierrez JA, Solenberg PJ, Pfluger PT, Czyzyk TA, Willency JA, Schurmann A, Joost HG, Jandacek RJ, Hale JE, Heiman ML, Tschop MH (2009) GOAT links dietary lipids with the endocrine control of energy balance. Nat Med 15(7):741–745

Kluge M, Schussler P, Schmidt D, Uhr M, Steiger A (2012) Ghrelin suppresses secretion of luteinizing hormone (LH) and follicle-stimulating hormone (FSH) in women. J Clin Endocrinol Metab 97(3):E448–E451

Kohno D, Gao HZ, Muroya S, Kikuyama S, Yada T (2003) Ghrelin directly interacts with neuropeptide-Y-containing neurons in the rat arcuate nucleus: Ca2+ signaling via protein kinase A and N-type channel-dependent mechanisms and cross-talk with leptin and orexin. Diabetes 52(4):948–956

Kohno D, Nakata M, Maekawa F, Fujiwara K, Maejima Y, Kuramochi M, Shimazaki T, Okano H, Onaka T, Yada T (2007) Leptin suppresses ghrelin-induced activation of neuropeptide Y neurons in the arcuate nucleus via phosphatidylinositol 3-kinase- and phosphodiesterase 3-mediated pathway. Endocrinology 148(5):2251–2263

Kohno D, Sone H, Minokoshi Y, Yada T (2008) Ghrelin raises [Ca2+]i via AMPK in hypothalamic arcuate nucleus NPY neurons. Biochem Biophys Res Commun 366(2):388–392

Kola B, Hubina E, Tucci SA, Kirkham TC, Garcia EA, Mitchell SE, Williams LM, Hawley SA, Hardie DG, Grossman AB, Korbonits M (2005) Cannabinoids and ghrelin have both central and peripheral metabolic and cardiac effects via AMP-activated protein kinase. J Biol Chem 280(26):25196–25201

Lamont EW, Patterson Z, Rodrigues T, Vallejos O, Blum ID, Abizaid A (2012) Ghrelin-deficient mice have fewer orexin cells and reduced cFOS expression in the mesolimbic dopamine pathway under a restricted feeding paradigm. Neuroscience 218:12–19

Loos RJ, Lindgren CM, Li S, Wheeler E, Zhao JH, Prokopenko I, Inouye M, Freathy RM, Attwood AP, Beckmann JS, Berndt SI, Jacobs KB, Chanock SJ, Hayes RB, Bergmann S, Bennett AJ, Bingham SA, Bochud M, Brown M, Cauchi S, Connell JM, Cooper C, Smith GD, Day I, Dina C, De S, Dermitzakis ET, Doney AS, Elliott KS, Elliott P, Evans DM, Sadaf Farooqi I, Froguel P, Ghori J, Groves CJ, Gwilliam R, Hadley D, Hall AS, Hattersley AT, Hebebrand J, Heid IM, Lamina C, Gieger C, Illig T, Meitinger T, Wichmann HE, Herrera B, Hinney A, Hunt SE, Jarvelin MR, Johnson T, Jolley JD, Karpe F, Keniry A, Khaw KT, Luben RN, Mangino M, Marchini J, McArdle WL, McGinnis R, Meyre D, Munroe PB, Morris AD, Ness AR, Neville MJ, Nica AC, Ong KK, O'Rahilly S, Owen KR, Palmer CN, Papadakis K, Potter S, Pouta A, Qi L, Randall JC, Rayner NW, Ring SM, Sandhu MS, Scherag A, Sims MA, Song K, Soranzo N, Speliotes EK, Syddall HE, Teichmann SA, Timpson NJ, Tobias JH, Uda M, Vogel CI, Wallace C, Waterworth DM, Weedon MN, Willer CJ, Wraight, Yuan X, Zeggini E, Hirschhorn JN, Strachan DP, Ouwehand WH, Caulfield MJ, Samani NJ, Frayling TM, Vollenweider P, Waeber G, Mooser V, Deloukas P, McCarthy MI, Wareham NJ, Barroso I, Jacobs KB, Chanock SJ, Hayes RB, Lamina C, Gieger C, Illig T, Meitinger T, Wichmann HE, Kraft P, Hankinson SE, Hunter DJ, Hu FB, Lyon HN, Voight BF, Ridderstrale M, Groop L, Scheet P, Sanna S, Abecasis GR, Albai G, Nagaraja R, Schlessinger D, Jackson AU, Tuomilehto J, Collins FS, Boehnke M, Mohlke KL (2008) Common variants near MC4R are associated with fat mass, weight and risk of obesity. Nature genetics 40(6):768–775

Lopez M, Lage R, Saha AK, Perez-Tilve D, Vazquez MJ, Varela L, Sangiao-Alvarellos S, Tovar S, Raghay K, Rodriguez-Cuenca S, Deoliveira RM, Castaneda T, Datta R, Dong JZ, Culler M, Sleeman MW, Alvarez CV, Gallego R, Lelliott CJ, Carling D, Tschop MH, Dieguez C, Vidal-

Puig A (2008) Hypothalamic fatty acid metabolism mediates the orexigenic action of ghrelin. Cell Metab 7(5):389–399

Luquet S, Perez FA, Hnasko TS, Palmiter RD (2005) NPY/AgRP neurons are essential for feeding in adult mice but can be ablated in neonates. Science 310(5748):683–685

Luquet S, Phillips CT, Palmiter RD (2007) NPY/AgRP neurons are not essential for feeding responses to glucoprivation. Peptides 28(2):214–225

Minokoshi Y, Alquier T, Furukawa N, Kim YB, Lee A, Xue B, Mu J, Foufelle F, Ferre P, Birnbaum MJ, Stuck BJ, Kahn BB (2004) AMP-kinase regulates food intake by responding to hormonal and nutrient signals in the hypothalamus. Nature 428(6982):569–574

Murphy BA, Fioramonti X, Jochnowitz N, Fakira K, Gagen K, Contie S, Lorsignol A, Penicaud L, Martin WJ, Routh VH (2009) Fasting enhances the response of arcuate neuropeptide Y-glucose-inhibited neurons to decreased extracellular glucose. Am J Physiol Cell Physiol 296(4):C746–C756

Nakazato M, Murakami N, Date Y, Kojima M, Matsuo H, Kangawa K, Matsukura S (2001) A role for ghrelin in the central regulation of feeding. Nature 409(6817):194–198

Naleid AM, Grace MK, Cummings DE, Levine AS (2005) Ghrelin induces feeding in the mesolimbic reward pathway between the ventral tegmental area and the nucleus accumbens. Peptides 26(11):2274–2279

O'Rahilly S, Farooqi IS (2008) Human obesity as a heritable disorder of the central control of energy balance. Int J Obes (Lond) 32(Suppl 7):S55–S61

Obici S, Feng Z, Arduini A, Conti R, Rossetti L (2003) Inhibition of hypothalamic carnitine palmitoyltransferase-1 decreases food intake and glucose production. Nat Med 9(6):756–761

Ogata R, Matsuzaki T, Iwasa T, Kiyokawa M, Tanaka N, Kuwahara A, Yasui T, Irahara M (2009) Hypothalamic Ghrelin suppresses pulsatile secretion of luteinizing hormone via beta-endorphin in ovariectomized rats. Neuroendocrinology 90(4):364–370

Olszewski PK, Grace MK, Billington CJ, Levine AS (2003) Hypothalamic paraventricular injections of ghrelin: effect on feeding and c-Fos immunoreactivity. Peptides 24(6):919–923

Perello M, Sakata I, Birnbaum S, Chuang JC, Osborne-Lawrence S, Rovinsky SA, Woloszyn J, Yanagisawa M, Lutter M, Zigman JM (2010) Ghrelin increases the rewarding value of high-fat diet in an orexin-dependent manner. Biol Psychiatry 67(9):880–886

Pinto S, Roseberry AG, Liu H, Diano S, Shanabrough M, Cai X, Friedman JM, Horvath TL (2004) Rapid rewiring of arcuate nucleus feeding circuits by leptin.[see comment]. Science 304(5667):110–115

Ramirez S, Martins L, Jacas J, Carrasco P, Pozo M, Clotet J, Serra D, Hegardt FG, Dieguez C, Lopez M, Casals N (2013) Hypothalamic ceramide levels regulated by CPT1C mediate the orexigenic effect of ghrelin. Diabetes 62(7):2329–2337

Reichenbach A, Steyn FJ, Sleeman MW, Andrews ZB (2012) Ghrelin receptor expression and colocalization with anterior pituitary hormones using a GHSR-GFP mouse line. Endocrinology 153(11):5452–5466

Sakata I, Yang J, Lee CE, Osborne-Lawrence S, Rovinsky SA, Elmquist JK, Zigman JM (2009) Colocalization of ghrelin O-acyltransferase and ghrelin in gastric mucosal cells. Am J Physiol Endocrinol Metab 297(1):E134–E141

Shrestha YB, Wickwire K, Giraudo S (2009) Effect of reducing hypothalamic ghrelin receptor gene expression on energy balance. Peptides 30(7):1336–1341

Smith JT, Reichenbach A, Lemus M, Mani BK, Zigman JM, Andrews ZB (2013) An eGFP-expressing subpopulation of growth hormone secretagogue receptor cells are distinct from kisspeptin, tyrosine hydroxylase, and RFamide-related peptide neurons in mice. Peptides 47:45–53

Spencer SJ, Xu L, Clarke MA, Lemus M, Reichenbach A, Geenen B, Kozicz T, Andrews ZB (2012) Ghrelin regulates the hypothalamic-pituitary-adrenal axis and restricts anxiety after acute stress. Biol psychiatry 72(6):457–465

Steinberg GR, Kemp BE (2009) AMPK in health and disease. Physiol Rev 89(3):1025–1078

Sternson SM, Shepherd GM, Friedman JM (2005) Topographic mapping of VMH –> arcuate nucleus microcircuits and their reorganization by fasting. Nat Neurosci 8(10):1356–1363

Tong Q, Ye C, McCrimmon RJ, Dhillon H, Choi B, Kramer MD, Yu J, Yang Z, Christiansen LM, Lee CE, Choi CS, Zigman JM, Shulman GI, Sherwin RS, Elmquist JK, Lowell BB (2007) Synaptic glutamate release by ventromedial hypothalamic neurons is part of the neurocircuitry that prevents hypoglycemia. Cell Metab 5(5):383–393

Tong Q, Ye CP, Jones JE, Elmquist JK, Lowell BB (2008) Synaptic release of GABA by AgRP neurons is required for normal regulation of energy balance. Nat Neurosci 11:998–1000

Toshinai K, Date Y, Murakami N, Shimada M, Mondal MS, Shimbara T, Guan JL, Wang QP, Funahashi H, Sakurai T, Shioda S, Matsukura S, Kangawa K, Nakazato M (2003) Ghrelin-induced food intake is mediated via the orexin pathway. Endocrinology 144(4):1506–1512

Tschop M, Smiley DL, Heiman ML (2000) Ghrelin induces adiposity in rodents. Nature 407(6806):908–913

Velasquez DA, Martinez G, Romero A, Vazquez MJ, Boit KD, Dopeso-Reyes IG, Lopez M, Vidal A, Nogueiras R, Dieguez C (2011) The central Sirtuin 1/p53 pathway is essential for the orexigenic action of ghrelin. Diabetes 60(4):1177–1185

Vousden KH, Ryan KM (2009) p53 and metabolism. Nat Rev Cancer 9(10):691–700

Wang L, Saint-Pierre DH, Tache Y (2002) Peripheral ghrelin selectively increases Fos expression in neuropeptide Y—synthesizing neurons in mouse hypothalamic arcuate nucleus. Neurosci Lett 325(1):47–51

Watterson KR, Bestow D, Gallagher J, Hamilton DL, Ashford FB, Meakin PJ, Ashford ML (2013) Anorexigenic and orexigenic hormone modulation of mammalian target of rapamycin complex 1 activity and the regulation of hypothalamic agouti-related protein mRNA expression. Neurosignals 21(1–2):28–41

Willer CJ, Speliotes EK, Loos RJ, Li S, Lindgren CM, Heid IM, Berndt SI, Elliott AL, Jackson AU, Lamina C, Lettre G, Lim N, Lyon HN, McCarroll SA, Papadakis K, Qi L, Randall JC, Roccasecca RM, Sanna S, Scheet P, Weedon MN, Wheeler E, Zhao JH, Jacobs LC, Prokopenko I, Soranzo N, Tanaka T, Timpson NJ, Almgren P, Bennett A, Bergman RN, Bingham SA, Bonnycastle LL, Brown M, Burtt NP, Chines P, Coin L, Collins FS, Connell JM, Cooper C, Smith GD, Dennison EM, Deodhar P, Elliott P, Erdos MR, Estrada K, Evans DM, Gianniny L, Gieger C, Gillson CJ, Guiducci C, Hackett R, Hadley D, Hall AS, Havulinna AS, Hebebrand J, Hofman A, Isomaa B, Jacobs KB, Johnson T, Jousilahti P, Jovanovic Z, Khaw KT, Kraft P, Kuokkanen M, Kuusisto J, Laitinen J, Lakatta EG, Luan J, Luben RN, Mangino M, McArdle WL, Meitinger T, Mulas A, Munroe PB, Narisu N, Ness AR, Northstone K, O'Rahilly S, Purmann C, Rees MG, Ridderstrale M, Ring SM, Rivadeneira F, Ruokonen A, Sandhu MS, Saramies J, Scott LJ, Scuteri A, Silander K, Sims MA, Song K, Stephens J, Stevens S, Stringham HM, Tung YC, Valle TT, Van Duijn CM, Vimaleswaran KS, Vollenweider P, Waeber G, Wallace C, Watanabe RM, Waterworth DM, Watkins N, Witteman JC, Zeggini E, Zhai G, Zillikens MC, Altshuler D, Caulfield MJ, Chanock SJ, Farooqi IS, Ferrucci L, Guralnik JM, Hattersley AT, Hu FB, Jarvelin MR, Laakso M, Mooser V, Ong KK, Ouwehand WH, Salomaa V, Samani NJ, Spector TD, Tuomi T, Tuomilehto J, Uda M, Uitterlinden AG, Wareham NJ, Deloukas P, Frayling TM, Groop LC, Hayes RB, Hunter DJ, Mohlke KL, Peltonen L, Schlessinger D, Strachan DP, Wichmann HE, McCarthy MI, Boehnke M, Barroso I, Abecasis GR, Hirschhorn JN (2009) Six new loci associated with body mass index highlight a neuronal influence on body weight regulation. Nat Genet 41(1):25–34

Willesen MG, Kristensen P, Romer J (1999) Co-localization of growth hormone secretagogue receptor and NPY mRNA in the arcuate nucleus of the rat. Neuroendocrinology 70(5):306–316

Wolfgang MJ, Cha SH, Sidhaye A, Chohnan S, Cline G, Shulman GI, Lane MD (2007) Regulation of hypothalamic malonyl-CoA by central glucose and leptin. Proc Natl Acad Sci USA 104(49):19285–19290

Wolfgang MJ, Kurama T, Dai Y, Suwa A, Asaumi M, Matsumoto S, Cha SH, Shimokawa T, Lane MD (2006) The brain-specific carnitine palmitoyltransferase-1c regulates energy homeostasis. Proc Natl Acad Sci USA 103(19):7282–7287

Wolfgang MJ, Lane MD (2006) The role of hypothalamic malonyl-CoA in energy homeostasis. J Biol Chem 281(49):37265–37269

Woods A, Dickerson K, Heath R, Hong SP, Momcilovic M, Johnstone SR, Carlson M, Carling D (2005) Ca2+/calmodulin-dependent protein kinase kinase-beta acts upstream of AMP-activated protein kinase in mammalian cells. Cell Metab 2(1):21–33

Yang J, Brown MS, Liang G, Grishin NV, Goldstein JL (2008) Identification of the acyltransferase that octanoylates ghrelin, an appetite-stimulating peptide hormone. Cell 132(3):387–396

Yang Y, Atasoy D, Su HH, Sternson SM (2011) Hunger states switch a flip-flop memory circuit via a synaptic AMPK-dependent positive feedback loop. Cell 146(6):992–1003

Yin W, Mu J, Birnbaum MJ (2003) Role of AMP-activated protein kinase in cyclic AMP-dependent lipolysis In 3T3-L1 adipocytes. J Biol Chem 278(44):43074–43080

Zhang CY, Baffy G, Perret P, Krauss S, Peroni O, Grujic D, Hagen T, Vidal-Puig AJ, Boss O, Kim YB, Zheng XX, Wheeler MB, Shulman GI, Chan CB, Lowell BB (2001) Uncoupling protein-2 negatively regulates insulin secretion and is a major link between obesity, beta cell dysfunction, and type 2 diabetes. Cell 105(6):745–755

Zhu X, Cao Y, Voogd K, Steiner DF (2006) On the processing of proghrelin to ghrelin. J Biol Chem 281(50):38867–38870

Zigman JM, Jones JE, Lee CE, Saper CB, Elmquist JK (2006) Expression of ghrelin receptor mRNA in the rat and the mouse brain. J Comp Neurol 494(3):528–548

The Vagus Nerve and Ghrelin Function

Yukari Date

Abstract Ghrelin, a gastrointestinal hormone originally discovered in human and rat stomach, functions as the only orexigenic signal produced by peripheral tissues. Although ghrelin is considered to affect hypothalamic neurons producing agouti-related protein (AgRP) and neuropeptide Y (NPY) and induce food intake, it is still unclear how peripherally administered ghrelin activates these neurons. The vagal afferent fibers are the major neuroanatomical linkage between the gastrointestinal tract and the nucleus tractus solitarii. Recently, several gastrointestinal hormones have been shown to transmit orexigenic or anorectic signals to the brain at least in part via the vagal afferent system. Indeed, blockade of the vagal afferent pathway abolishes ghrelin-induced feeding, indicating that the vagal afferent system is important to convey orexigenic ghrelin signals to the brain. In this chapter, we mention the role of the vagal afferent system for feeding regulation by gastrointestinal hormones and show the functional linkage in feeding between peripheral ghrelin and the vagal afferent system.

Keywords Vagus nerve · Nodose ganglion · Nucleus tractus solitarii (NTS) · Gastrointestinal hormones · Orexigenic signals · Anorectic signals

Anatomical Characteristics of the Vagus Nerve

The vagus nerve is a complex nerve that innervates nearly all of the thoracic and abdominal viscera. It transmits information to and from the viscera as well as to and from cranial structures. The vagus nerve contains a diverse population of nerve fibers, such as those involved in visceromotor or viscerosensory functions.

Y. Date (✉)
Frontier Science Research Center, University of Miyazaki, Miyazaki 889-1692, Japan
e-mail: dateyuka@med.miyazaki-u.ac.jp

J. Portelli and I. Smolders (eds.), *Central Functions of the Ghrelin Receptor*,
The Receptors 25, DOI: 10.1007/978-1-4939-0823-3_4,
© Springer Science+Business Media New York 2014

Visceromotor functions consist of parasympathetic outflow from the medulla oblongata, whereas viscerosensory innervation includes afferent information from the gastrointestinal tract and cardiorespiratory axis. Although the sensory fibers innervating the digestive tract are intermingled with efferent fibers, the proportion of sensory fibers exceeds that of efferent fibers. Indeed, approximately 90 % of the vagus nerve in the subdiaphragm is afferent (Agostoni et al. 1957), indicating that this nerve is an important pathway for conveying information from the gastrointestinal tract directly to the brain. Thus, the vagus nerve forms a neuroanatomical linkage between the gastrointestinal tract and the brain.

The cell bodies of vagal afferent nerve fibers are located within the nodose ganglion, which is a prominent swelling of the vagus nerve emerging from the cranial cavity through the jugular foramen. There are about 6,000 neurons in the nodose ganglion of a rat, and the full set of ganglion neurons is already present at birth (Cooper 1984). The vagal afferent fibers innervating the abdominal viscera originate from vagal afferent neurons present in the body and caudal pole of the nodose ganglion (Dockray and Sharkey 1986). Substance P and calcitonin gene-related peptide have been recognized as suitable markers for vagal afferent neurons and terminals (Green and Dockray 1987). Central branches of the nodose ganglion terminate in the nucleus tractus solitarius (NTS) of the medulla oblongata, the first synaptic site for the afferent projections of the vagus nerve from the gastrointestinal tract. The nerve fibers ascending from the NTS reach a number of forebrain sites, including the hypothalamic nuclei involved in the regulation of feeding.

The Role of the Vagus Nerve in Gastrointestinal Hormone Action

Several sensory signals from the gastrointestinal tract that are involved in feeding behavior are delivered to the NTS primarily by the vagal afferents that terminate centrally within the caudal nucleus of the NTS (Rinaman 2010). These signals are subsequently relayed to the forebrain by monoaminergic and/or peptidergic projections arising within the NTS. In addition, some vagal afferent fibers terminate within the gastrointestinal mucosa and submucosa and are optimally positioned to monitor the composition of the gastrointestinal lumen or the concentration of bioactive substances released from enteroendocrine cells (Grundy and Scratcherd 1989). Several gastrointestinal hormones influence the feeding circuit in the central nervous system at least in part via the vagal afferent pathway; these include cholecystokinin (CCK), peptide YY (PYY), and glucagon-like peptide-1 (GLP-1), which function as satiety signals.

CCK is produced in the enteroendocrine cells that are distributed evenly throughout the duodenum and proximal jejunum; these cells can directly access nutrients (Walsh 1987). In rats, intragastric administration of a mixed meal increases plasma CCK levels (Liddle et al. 1986). Administration of individual nutrients showed that intact protein stimulates CCK release, whereas administration of hydrolyzed protein, amino acids, starch, or fat does not (Raybould 1992). Thus, CCK appears to be released in response to intact protein and to function as a satiety signal by inhibiting gastric emptying (Green et al. 1989). CCK receptors, G protein-coupled receptors, consist of two different types: CCK-A receptor and CCK-B receptor (Want 1995). CCK-A receptor has high affinity to sulfated CCK which is known as a bioactive CCK and is expressed in vagal afferent neurons as well as gallbladder. CCK-A receptor produced in vagal afferent neurons is transported to the periphery (Zarbin et al. 1981; Date et al. 2005). CCK binding sites in the periphery are present on all subdiaphragmatic branches of the vagus nerve, and bound CCK-8 (s; sulfated) is not displaced by desulfated CCK (Moran et al. 1987). These findings indicate that the CCK receptor on vagal afferents is the peripheral subtype A (CCK-A receptor). Peripheral administration of CCK to rats reduces food intake, but bilateral subdiaphragmatic vagotomy abolishes the satiety effect of CCK (Smith et al. 1981). These data indicate that the vagus nerve plays an important role in transmitting CCK signals to the brain.

PYY, a gastrointestinal hormone that is produced in enteroendocrine cells of the ileum (Ekblad and Sundler 2002; Böttcher et al. 1986; Greeley et al. 1989), belongs to the neuropeptide Y (NPY) peptide family. Although NPY increases food intake, PYY reduces food intake by binding to the NPY Y2 receptor (Y2-R; Batterham et al. 2002). PYY is released in response to both neural and humoral factors and luminal nutrient content (Adrian et al. 1985). Because the Y2-R is mainly present in the hypothalamus, PYY has been thought to reduce feeding via the Y2-R located in the hypothalamus. However, Y2-R is also produced in the vagal afferent neurons and transported to the vagal afferent termini (Abbott et al. 2005; Koda et al. 2005). Considering that PYY is produced in the enteroendocrine cells and its receptor is present in the vagus nerve, the satiety signal of PYY appears to be conveyed to the NTS partially via the vagal afferent pathway. Indeed, peripheral administration of PYY to vagotomized rats does not reduce feeding (Abbott et al. 2005; Koda et al. 2005). Furthermore, a single administration of PYY to sham-operated rats induces the cFos protein, a marker of neuronal activation, in some neurons of the hypothalamic arcuate nucleus (ARC), whereas cFos expression is attenuated when PYY is administered to vagotomized rats (Koda et al. 2005). These data suggest the possibility that the PYY signal is also transmitted to the hypothalamus at least in part via the vagal afferent pathway.

GLP-1, a gastrointestinal hormone, is released by the enteroendocrine L cells in the small intestine in response to nutrients (Holst et al. 2007). This hormone is also produced in the NTS, and GLP-1 receptors (GLP-1-R) are expressed in a variety of peripheral tissues as well as the brain (Holst et al. 2007; Hayes et al. 2010). GLP-1 injected into rats either peripherally or centrally reduces feeding. GLP-1 has been reported to cross the blood–brain barrier and reduce feeding via its receptor (Kastin et al. 2002), which is present in the brain; however, it has also been speculated that only small amounts of active GLP-1 can pass the blood–brain barrier. Baggio et al. indicated that Albugon, a fusion protein of albumin and GLP-1, cannot pass the blood–brain barrier, although it can reduce feeding (Baggio et al. 2004). This finding suggests that GLP-1 reduces feeding via its receptor, GLP-1-R, which is expressed in the nodose ganglion. Supporting this, the GLP-1-induced reduction of feeding is attenuated in vagotomized rats, and peripherally administered GLP-1 reduces feeding even after central administration of a GLP-1 receptor antagonist. Peripheral administration of GLP-1, as well as CCK and PYY, increases the firing rate of the vagal afferent fibers (Date et al. 2005; Koda et al. 2005; Nakabayashi et al. 1996). It appears that GLP-1 produced in the gastrointestinal tract binds to its receptor on the vagal afferents and alters the firing rate of the afferent fibers, thereby functioning as a satiety signal.

The Role of the Vagus Nerve in Ghrelin-Induced Feeding

Ghrelin and the Ghrelin Receptor

Ghrelin, a gastrointestinal hormone primarily produced in the stomach, functions in feeding control as well as in growth hormone secretion by binding to the growth hormone secretagogue receptor (ghrelin receptor) (Kojima et al. 1999; Nakazato et al. 2001; Tschöp et al. 2001; Wren et al. 2000). In contrast to CCK, GLP-1, and PYY, which all contribute to reductions in feeding, ghrelin is the only gastrointestinal hormone known to accelerate feeding. Indeed, peripherally administered ghrelin markedly increases food intake (Tschöp et al. 2001; Date et al. 2002). Plasma ghrelin levels increase before and decrease after meals (Shiiya et al. 2000). These findings indicate that ghrelin acts as a starvation signal.

Ghrelin-producing cells are present in the oxyntic glands of the stomach (Date et al. 2000) and colocalize with chromogranin A–immunoreactive cells, which suggests that ghrelin is produced by endocrine cells in the stomach. Immunoelectron microscopy has identified the morphological characteristics of ghrelin-containing granules as round, compact, and electron dense. Because of the similarities in ultrastructural features between ghrelin and X/A-like cells, which make up 20 % of endocrine cells in the oxyntic gland, ghrelin cells are thought to be X/A-like cells. Very recently, Gagnon and Anini (2012) successfully created a primary culture of ghrelin-producing cells. By using this culture system, they found that noradrenaline (NAD) stimulates ghrelin secretion through the β1-adrenergic receptors on ghrelin cells, whereas insulin inhibits ghrelin secretion via the insulin receptor α- and β-subunits (Gagnon and Anini 2012). It has been reported that disruption of vagus nerve decreases ghrelin mRNA and inclines active ghrelin levels (Erlanson-Albertsson and Lindqvist 2008). Furthermore, Takiguchi et al. showed that preservation of the vagus nerve during laparoscopy-associated gastrectomy decreased postprandial ghrelin levels compared to not-preservation of it (Takiguchi et al. 2013). Thus, the autonomic nervous system is at least partially involved in the ghrelin secretion.

Ghrelin receptor, which is a G protein-coupled receptor specific for ghrelin, was identified as a receptor for small synthetic molecules (GHSs) that induce growth hormone secretion from the pituitary. Before the discovery of ghrelin, ghrelin receptor was considered to be an orphan G protein-coupled receptor. The ghrelin receptor is mainly present in the pituitary, hypothalamus, and hippocampus. In addition, it has been detected in the pancreas, adipose tissue, immune cells, cardiovascular system, and nodose ganglion (Laviano et al. 2012; Date et al. 2002).

Appetite Control by Ghrelin

Ghrelin is the only starvation hormone produced by peripheral tissues. Intravenous or subcutaneous administration of ghrelin significantly increases food intake (Nakazato et al. 2001; Tschöp et al. 2001). Peripherally administered ghrelin also stimulates NPY- and agouti-related protein (AgRP)-producing neurons. Given that the ghrelin receptor is present on NPY- and AgRP-producing neurons located in the hypothalamic ARC (Mondal et al. 2005), ghrelin may cross the blood–brain barrier to activate NPY and AgRP, thereby inducing feeding. However, whether peripheral ghrelin can cross the blood–brain barrier is controversial. Recently, it was shown that plasma ghrelin crosses the blood–brain barrier at a fairly low rate (Fry et al. 2010). If so, peripheral ghrelin must stimulate the appropriate hypothalamic regions via an indirect pathway, such as the vagal afferent pathway. To elucidate the role of the vagus nerve in ghrelin-induced feeding, we investigated food intake after ghrelin administration to rats that had undergone bilateral subdiaphragmatic vagotomy. A single dose of ghrelin significantly increased the 2 h food intake of sham-operated rats, but did not increase food intake by rats that had undergone subdiaphragmatic vagotomy (Date et al. 2002). Peripheral administration of ghrelin induces the cFos protein in some neurons in the ARC of rat hypothalamus; however, ghrelin did not induce cFos in the vagotomized rats. Ghrelin receptors produced in the nodose ganglion are transported to the vagus afferent termini through axonal flow, and peripherally administered ghrelin significantly decreases the firing rate of the afferent fibers (Date et al. 2002). These findings indicate that the vagal afferents play an important role in mediating the ghrelin starvation signal.

Central Integration of Ghrelin Signals in the Brain

The NTS receives information via the vagal afferent pathway. After ghrelin administration to rats, mRNA expression of dopamine β-hydroxylase (an NA synthetic enzyme) increases in the NTS (Date et al. 2006). NAD-producing neurons are a major population of the NTS, and NAD that originates from the NTS projects to NPY neurons in the hypothalamic ARC. Thus, peripherally administered ghrelin increases the NAD concentration in the ARC (Date et al. 2006). Furthermore, bilateral midbrain transections rostral to the NTS abolish ghrelin-induced feeding (Date et al. 2006). These findings indicate that the noradrenergic pathway from the NTS to the hypothalamus is necessary in the central control of the sensation of starvation transmitted by peripheral ghrelin.

AMP-activated protein kinase (AMPK) is involved in hypothalamic regulation of feeding (Minokoshi et al. 2004; Poleni et al. 2012); for example, leptin decreases hypothalamic AMPK activity, suppressing feeding (Mimokoshi et al. 2004). We found that coadministration of GLP-1 with leptin at subthreshold levels significantly decreases both AMPK activity in the hypothalamus of rats and their food intake (Poleni et al. 2012). These findings indicate that metabolic system through AMPK in the hypothalamus is crucial for the regulation of feeding. In contrast to the effects of leptin, ghrelin administration increases AMPK activity in the hypothalamus (Andersson et al. 2004). As mentioned earlier, peripheral ghrelin increases NAD release in the hypothalamic ARC via the NTS. In addition, NAD-containing fibers innervate NPY-producing neurons. Information about feeding that is integrated in the NTS or directly reaches the hypothalamus could therefore regulate energy homeostasis at least in part through the regulation of metabolic system via hypothalamic AMPK.

Conclusion

We and other research groups have shown that humoral signals related to feeding are transmitted to the brain at least in part through the vagal afferent pathway. Considering the anatomical and functional relationship between gastrointestinal hormones and the vagus nerve, it is plausible that the vagal afferent pathway is a major route via which peripheral orexigenic, anorectic, and/or other signals are conveyed to the brain. Indeed, the vagal afferent pathway plays an important role in transmitting information of not only feeding but also inflammation produced by peripheral substances including ghrelin (Rosas-Ballina and Tracey 2009; Baatar et al. 2011). Feeding is finely regulated by the complicated interaction of many factors produced in the peripheral tissues and brain. Ghrelin is also known to be produced in the brain. Ghrelin receptor is abundantly present in the hippocampus. Taken together, ghrelin/its receptor system may contribute not only to the initiation of a meal but also to searching food on the basis of memory retention (Olszewski et al. 2008). Further investigation of the mechanisms of the interactions and/or integration of feeding-related hormones in the periphery and brain will increase our understanding of the physiological roles of these hormones in feeding and energy homeostasis.

Acknowledgments We thank A. Niijima, N. Murakami, M. Nakazato, and K. Kangawa for their technical advice and helpful discussions. This work was supported, in part, by grants-in-aid from the Ministry of Education, Culture, Sports, Science, and Technology of Japan, and the Program for the Promotion of Basic Research Activities for Innovative Bioscience (PROBRAIN).

References

Abbott CR, Monteiro M, Small CJ et al (2005) The inhibitory effects of peripheral administration of peptide YY (3–36) and glucagon-like peptide-1 on food intake are attenuated by ablation of the vagal-brainstem-hypothalamic pathway. Brain Res 1044:127–131

Adrian TE, Ferri GL, Bacarese-Hamilton AJ et al (1985) Human distribution and release of a putative new gut hormone, peptide YY. Gastroenterology 89:1070–1077

Agostoni E, Chinnock JE, de Daly MB et al (1957) Functional and histological studies of the vagus nerve and its branches to the heart, lungs and abdominal viscera in the cat. J Physiol 135:182–205

Andersson U, Filipsson K, Abbott CR et al (2004) AMP-activated protein kinase plays a role in the control of food intake. J Biol Chem 279:12005–12008

Baatar D, Patel K, Taub DD (2011) The effects of ghrelin on inflammation and the immune system. Mol Cell Endocrinol 340:44–58

Baggio LL, Huang Q, Brown TJ et al (2004) A recombinant human glucagon-like peptide (GLP)-1-albumin protein (albugon) mimics peptidergic activation of GLP-1 receptor-dependent pathways coupled with satiety, gastrointestinal motility, and glucose homeostasis. Diabetes 53:2492–2500

Batterham RL, Cowley MA, Small CJ et al (2002) Gut hormone PYY(3-36) physiologically inhibits food intake. Nature 418:650–654

Böttcher G, Alumets J, Håkanson R et al (1986) Co-existence of glicentin and peptide YY in colorectal L-cells in cat and man. An electron microscopic study. Regul Pept 13:283–291

Cooper E (1984) Synapse formation among developing sensory neurons from rat nodose ganglia grown in tissue culture. J Physiol 351:263–274

Date Y, Kojima M, Hosoda H et al (2000) Ghrelin, a novel growth hormone-releasing acylated peptide, is synthesized in a distinct endocrine cell type in the gastrointestinal tracts of rats and humans. Endocrinology 141:4255–4261

Date Y, Murakami N, Toshinai K et al (2002) The role of the gastric afferent vagal nerve in ghrelin-induced feeding and growth hormone secretion in rats. Gastroenterology 123:1120–1128

Date Y, Toshinai K, Koda S et al (2005) Peripheral interaction of ghrelin with cholecystokinin on feeding regulation. Endocrinology 146:3518–3525

Date Y, Shimbara T, Koda S et al (2006) Peripheral ghrelin transmits orexigenic signals through the noradrenergic pathway from the hindbrain to the hypothalamus. Cell Metab 4:323–331

Dockray GJ, Sharkey KA (1986) Neurochemistry of visceral afferent neurons. Prog Brain Res 67:133–148

Ekblad E, Sundler F (2002) Distribution of pancreatic polypeptide and peptide YY. Peptides 23:251–261

Erlanson-Albertsson C, Lindqvist A (2008) Vagotomy and accompanying pyloroplasty down-regulates ghrelin mRNA but does not affect ghrelin secretion. Regul Pept 151:14–18

Fry M, Ferguson AV (2010) Ghrelin: central nervous system sites of action in regulation of energy balance. Int J Pept 2010:1–8

Gagnon J, Anini Y (2012) Insulin and norepinephrine regulate ghrelin secretion from a rat primary stomach cell culture. Endocrinology 153:3646–3656

Green T, Dockray GJ (1987) Calcitonin gene-related peptide and substance P in afferents to the upper gastrointestinal tract in the rat. Neurosci Lett 76:151–156

Green T, Dimaline R, Dockray GJ (1989) Neuroendocrine control mechanisms of gastric emptying in the rat. In: Singer MV, Goebell H (eds) Nerves and the gastrointestinal tract. Kluwer, Netherlands, pp 433–446

Greeley GHJ, Jeng YJ, Gomez G et al (1989) Evidence for regulation of peptide-YY release by the proximal gut. Endocrinology 124:1438–1443

Grundy D, Scratcherd T (1989) Sensory afferents from the gastrointestinal tract. In: Schultz SG (ed) Handbook of physiology: the gastrointestinal system. Motility and circulation, vol. 1. Oxford University Press, New York, pp 593–620

Hayes MR, De Jonghe BC, Kanoski SE (2010) Role of the glucagon-like-peptide-1 receptor in the control of energy balance. Physiol Behav 100:503–510

Holst JJ (2007) The physiology of glucagon-like peptide 1. Physiol Rev 87:1409–1439

Kastin AJ, Akerstrom V, Pan W (2002) Interactions of glucagon-like peptide-1 (GLP-1) with the blood–brain barrier. J Mol Neurosci 18:7–14

Koda S, Date Y, Murakami N et al (2005) The role of the vagal nerve in peripheral PYY_{3-36}-induced feeding reduction in rats. Endocrinology 146:2369–2375

Kojima M, Hosoda H, Date Y et al (1999) Ghrelin is a novel growth hormone releasing acylated peptide from stomach. Nature 402:656–660

Laviano A, Molfino A, Rianda S et al (2012) The growth hormone secretagogue receptor (Ghs-R). Curr Pharm Des 18:4749–4754

Liddle RA, Green GM, Conrad CK et al (1986) Proteins but not amino acids, carbohydrates, or fats stimulate cholecystokinin secretion in the rat. Am J Physiol 251:G243–G248

Minokoshi Y, Alquier T, Furukawa N et al (2004) AMP-kinase regulates food intake by responding to hormonal and nutrient signals in the hypothalamus. Nature 428:569–574

Mondal MS, Date Y, Yamaguchi H et al (2005) Identification of ghrelin and its receptor in neurons of the rat arcuate nucleus. Regul Pept 126:55–59

Moran TH, Smith GP, Hostetler AM et al (1987) Transport of cholecystokinin (CCK) binding sites in subdiaphragmatic vagal branches. Brain Res 415:149–152

Nakabayashi H, Nishizawa M, Nakagawa A et al (1996) Vagal hepatopancreatic reflex effect evoked by intraportal appearance of tGLP-1. Am J Physiol 271:E808–E813

Nakazato M, Murakami N, Date Y et al (2001) A role for ghrelin in the central regulation of feeding. Nature 409:194–198

Olszewski PK, Schiöth HB, Levine AS (2008) Ghrelin in the CNS: from hunger to a rewarding and memorable meal? Brain Res Rev 58:160–170

Poleni PE, Akieda-Asai S, Koda S et al (2012) Possible involvement of melanocortin-4-receptor and AMP-activated protein kinase in the interaction of glucagon-like peptide-1 and leptin on feeding in rats. Biochem Biophys Res Commun 420:36–41

Raybould HE (1992) Vagal afferent innervation and the regulation of gastric motor function. In: Ritter S, Ritter RC, Barnes CD (eds) Neuroanatomy and physiology of abdominal vagal afferents. CRC press, Florida, pp 193–219

Rinaman L (2010) Ascending projections from the caudal visceral nucleus of the solitary tract to brain regions involved in food intake and energy expenditure. Brain Res 1350:18–34

Rosas-Ballina M, Tracey KJ (2009) Cholinergic control of inflammation. J Intern Med 265:663–679

Shiiya T, Nakazato M, Mizuta M et al (2000) Plasma ghrelin levels in lean and obese humans and the effect of glucose on ghrelin secretion. J Clin Endocrinol Metab 87:240–244

Smith GP, Jerome C, Cushin BJ et al (1981) Abdominal vagotomy blocks the satiety effect of cholecystokinin in the rat. Science 213:1036–1037

Takiguchi S, Fujiwara Y, Yamasaki M et al (2013) Laparoscopy-assisted distal gastrectomy versus open distal gastrectomy. A prospective randomized single-blind study. World J Surg (in press)

Tschöp M, Weyer C, Tataranni PA et al (2001) Circulating ghrelin levels are decreased in human obesity. Diabetes 50:707–709

Walsh JH (1987) Gastrointestinal hormones. In: Johnson LR (ed) Physiology of the gastrointestinal tract. Raven Press, New York, pp 181–253

Want SA (1995) Cholecystokinin receptors. Am J Physiol 269:G628–G646

Wren AM, Small CJ, Ward HL et al (2000) The novel hypothalamic peptide ghrelin stimulates food intake and growth hormone secretion. Endocrinology 141:4325–4328

Zarbin MA, Wamsley JK, Innis RB et al (1981) Cholecystokinin receptors: presence and axonal flow in the rat vagus nerve. Life Sci 29:697–705

Part III
Ghrelin Receptors in Food and Drug Addictive Mechanisms

Central Ghrelin Receptors and Food Intake

Mario Perello and Jesica Raingo

Abstract Feeding is a vital function that provides nutritional and energy metabolism needs for animals. To ensure feeding, mammalian brains possess several interrelated neuronal systems that regulate different aspects of feeding behaviors. These neuronal circuits controlling food intake are strongly regulated by peripheral signals that contribute to the fine regulation of the energy homeostasis, such as metabolites and hormones. Among the signals regulating food intake, the stomach-derived hormone ghrelin and its receptor [named ghrelin receptor or the growth hormone secretagogue receptor type 1a (ghrelin receptor 1a)] play a major role. Ghrelin is the only mammalian peptide hormone able to increase food intake. Ghrelin stimulates appetite by affecting both food intake itself and also the rewarding aspects of feeding. As discussed below, the central distribution of ghrelin receptor 1a supports the concept that ghrelin regulates both homeostatic and hedonic aspects of feeding, and evidence from different studies confirms that ghrelin promotes food intake via diverse mechanisms. Of note, derangements in the ghrelin/ghrelin receptor 1a system have been reported in several eating disorders, including obesity, anorexia nervosa, bulimia nervosa, binge eating disorders, cachexia, and Prader-Willi syndrome. Here, the potential pathways by which ghrelin receptor 1a regulates feeding, with a special focus on hedonic aspects of eating, are delineated. Also, recent evidence suggesting a role of the ghrelin system in disorders with alterations of food intake is briefly reviewed.

Keywords Homeostatic eating · Hedonic eating · Food reward · Hypothalamus · Mesolimbic pathway

M. Perello (✉)
Laboratory of Neurophysiology, Multidisciplinary Institute of Cell Biology, Calle 526 S/N entre 10 y 11, PO Box 403, 1900 La Plata, Buenos Aires, Argentina
e-mail: marioperello@yahoo.com; mperello@imbice.gov.ar

J. Raingo
Laboratory of Electrophysiology of the Multidisciplinary Institute of Cell Biology (IMBICE), Argentine Research Council (CONICET) and Scientific Research Commission of the Province of Buenos Aires (CIC-PBA), La Plata, Buenos Aires, Argentina

J. Portelli and I. Smolders (eds.), *Central Functions of the Ghrelin Receptor*, The Receptors 25, DOI: 10.1007/978-1-4939-0823-3_5,
© Springer Science+Business Media New York 2014

Homeostatic and Hedonic Feeding Circuits

Feeding regulation involves an integrated regulatory system in which homeostatic brain circuits, that drive food intake depending on energy store levels, interact with the hedonic circuits that drive consumption based on rewarding properties of foods (Berthoud 2011; Saper et al. 2002). The homeostatic circuits provide a means by which signals of energy availability, including ghrelin, modulate food intake (Williams and Elmquist 2012; Schwartz et al. 2000). Thus, homeostatic-driven feeding occurs under negative energy balance conditions, when fuel stores are depleted and plasma ghrelin is elevated (Williams and Elmquist 2012; Schwartz et al. 2000). In contrast, hedonic-driven feeding refers to the involvement of cognitive, reward, and emotional factors that lead to the consumption of pleasurable foods even when extra calories are not necessary (Berthoud 2011; Saper et al. 2002). Neuronal systems controlling homeostatic feeding are located mainly in the brainstem and hypothalamus while neuronal systems controlling hedonic feeding are primarily related to cortico-limbic structures (Berthoud 2011; Saper et al. 2002; Williams and Elmquist 2012; Schwartz et al. 2000). Importantly, both homeostatic and hedonic brain circuits driving food intake are regulated by peripheral signals.

The hypothalamus contains several nuclei involved in food intake regulation, including the arcuate nucleus (ARC), the paraventricular nucleus (PVN), the lateral hypothalamic area (LHA), the ventromedial nucleus (VMN), and the dorsomedial nucleus (DMN) (Williams and Elmquist 2012; Schwartz et al. 2000; Suzuki et al. 2010). The ARC has become a major focus for energy balance research because circulating factors, such as ghrelin, have increased accessibility to this nucleus, where receptors for peripheral signals are highly expressed (Williams and Elmquist 2012; Schwartz et al. 2000; Suzuki et al. 2010). The ARC contains a key set of neurons that express the potent orexigenic neuropeptides agouti-gene-related protein (AgRP) and neuropeptide Y (NPY), and also the neurotransmitter γ-aminobutyric acid (GABA) (Williams and Elmquist 2012; Schwartz et al. 2000; Suzuki et al. 2010). To explain homeostatic food intake, initial emphasis has been placed on a simple model in which ARC neurons act as first-order neurons that sense peripheral factors and then regulate second-order neurons of the PVN, VMN, DMH, and LHA (Williams and Elmquist 2012; Schwartz et al. 2000; Suzuki et al. 2010). Recent evidence shows that another target of ARC neurons is the parabrachial nucleus (PBN), which is located in the hindbrain and inhibits feeding (Wu and Palmiter 2011; Atasoy et al. 2012). Second-order neurons project then to other brain areas, including the dorsal vagal complex in the brainstem, which comprises the nucleus tractus solitarius (NTS), the area postrema (AP), and the dorsomotor nucleus of the vagus (DMV), and plays a major role regulating food intake in concert with the ARC (Williams and Elmquist 2012; Schwartz et al. 2000; Suzuki et al. 2010). The dorsal vagal complex senses peripheral hormones directly and also integrates neuronal inputs from the hypothalamic and peripheral centers. In particular, the NTS is a termination site of the vagal afferent fibers that transmit visceral sensory information, including gastric distension and gut factors, from cell bodies located in the nodose ganglia (Williams and Elmquist 2012; Schwartz et al. 2000; Suzuki et al. 2010). Thus, homeostatic adjustments of food intake integrate not only hypothalamic systems governing intake on a meal-to-meal basis but also brainstem systems regulating meal size and/or meal frequency.

A key element of neuronal circuits regulating food reward behaviors is the dopaminegic pathway emanating from the midbrain ventral tegmental area (VTA) (Berthoud 2011; Saper et al. 2002; DiLeone et al. 2012; Hyman et al. 2006). Dopaminergic VTA neurons project to the nucleus accumbens (NAc) in the ventral striatum and other areas such as the amygdala, medial prefrontal cortex (mPFC), hippocampus, and hypothalamus (DiLeone et al. 2012; Hyman et al. 2006). The VTA receives projections from many brain nuclei, including the above-mentioned areas that receive projections from the VTA and cholinergic neurons of the laterodorsal tegmental area (LDTg) (Dickson et al. 2010). In addition, the VTA receives taste information via afferent sensory fibers that have two brainstem relays, in the NTS and in the PBN (DiLeone et al. 2012; Hyman et al. 2006). Dopamine release in the NAc potently augments the drive to obtain food rewards (Palmiter 2007). The shell part of the NAc is particularly important for eating behaviors since it sends projections to the LHA neurons controlling food intake (Stratford and Kelley 1999; Zheng et al. 2007). Orexigenic LHA neurons seem to be under a tonic inhibition that can be relieved by activation of reward pathways (Stratford and Kelley 1999; Zheng et al. 2007). In addition, LHA orexin neurons send projections to the VTA, where they activate dopaminergic neurons (Nakamura et al. 2000; Korotkova et al. 2003). Thus, LHA orexin neurons have been proposed as a potential link between homeostatic and hedonic circuits regulating food intake (Mahler et al. 2012).

Ghrelin and Ghrelin Receptor 1a in Feeding Centers

The ghrelin receptor 1a is present in and regulates both homeostatic and hedonic feeding centers (Perello and Zigman 2012; Skibicka and Dickson 2011; Zigman et al. 2006; Guan et al. 1997). Initially, ghrelin was shown to stimulate food intake by acting on homeostatic hypothalamic circuits (Nakazato et al. 2001; Briggs and Andrews 2011). Ghrelin effects on homeostatic eating likely involve the NPY/AgRP/GABA neurons of the ARC that express high levels of ghrelin receptor 1a (Nakazato et al. 2001; Briggs and Andrews 2011; Kageyama et al. 2010; Willesen et al. 1999). Ghrelin-induced food intake also seems to depend on orexin neurons of the LHA, where ghrelin receptor 1a is expressed (Toshinai et al. 2003; Olszewski et al. 2003). Additionally, some evidence indicates that the vagus nerve integrity is required for ghrelin-induced food intake (Date 2012; Date et al. 2002). According to this possibility, ghrelin receptor 1a is expressed in vagal afferent neurons of nodose ganglia and in the dorsal vagal complex (Zigman et al. 2006;

Sakata et al. 2003). The presence of ghrelin receptor 1a in dopaminergic VTA neurons supports the possibility that ghrelin can regulate hedonic aspects of eating (Abizaid et al. 2006; Zigman et al. 2006; Chuang et al. 2011). Ghrelin may also regulate mesolimbic circuits indirectly via the cholinergic neurons of the LDTg, which express ghrelin receptor 1a (Dickson et al. 2010; Jerlhag et al. 2008). Ghrelin's action on food reward requires intact orexin signaling; however, the neuronal circuits by which ghrelin recruits the LHA orexin neurons are still unknown (Perello et al. 2010). Ghrelin presumably affects eating behaviors by also acting on the hippocampus, a brain structure involved in memory and decision making that expresses ghrelin receptor 1a (Zigman et al. 2006; Diano et al. 2006). Figure 1 summarizes the ghrelin targets and the potential neuronal circuits controlling homeostatic and hedonic aspects of food intake affected by ghrelin.

The ability of ghrelin to act in the brain and increase food intake depends on the accessibility of circulating ghrelin to the above-mentioned brain areas. Circulating ghrelin cannot freely cross the blood–brain barrier, and it is currently unclear how this hormone enters the brain (Fry and Ferguson 2010). In mice, ghrelin can be transported from the brain for circulation via a saturable transport system; however, no such system has been identified for blood to brain transport (Banks 2008). It is frequently assumed that circulating ghrelin is able to access to the ARC, where blood–brain barrier is presumably weaker; however, this possibility is still under debate (Fry and Ferguson 2010; Rodriguez et al. 2010; Schaeffer et al. 2013). Another possibility is that circulating ghrelin gains access to the brain through the sensory circumventricular organs, which are specialized areas with fenestrated capillaries. The median eminence, located in close apposition to the ARC, is a circumventricular organ where plasma ghrelin can easily diffuse to reach neuronal ghrelin receptor 1a (Schaeffer et al. 2013). The AP is another circumventricular organ also known to participate in food intake regulation and that expresses ghrelin receptor 1a (Fry and Ferguson 2007, 2010; Zigman et al. 2006). Thus, circulating ghrelin could directly act on AP neurons, which then innervate several hypothalamic and brainstem feeding centers (Fry and Ferguson 2007). Some evidence does suggest that ghrelin-induced feeding depends on intact signaling at the AP (Gilg and Lutz 2006; Date et al. 2006).

The relevance of the expression of ghrelin receptor 1a in brain areas without access to circulating ghrelin is unclear. It has been proposed that ghrelin can be centrally produced; however, evidence about the source and physiological significance of centrally produced ghrelin is inconsistent (Cowley et al. 2003; Sakata et al. 2009; Furness et al. 2011). Ghrelin receptor 1a mainly signals through $G\alpha_{q/11}$, phospholipase C, inositol phosphate, and calcium mobilization from intracellular stores; although it also activates other signaling pathways (Cong et al. 2010). An interesting feature of ghrelin receptor 1a is its strong constitutive activity that makes it capable to signal in a ghrelin-independent manner (Mokrosinski and Holst 2010; Damian et al. 2012). Thus, the increase of ghrelin receptor 1a expression would accordingly increase activation of the downstream signaling pathways affecting, as a consequence, food intake and body weight regulation (Petersen et al. 2009). Additionally, it has been proposed that an alternative

Neuronal Circuits Controlling Homeostatic Eating

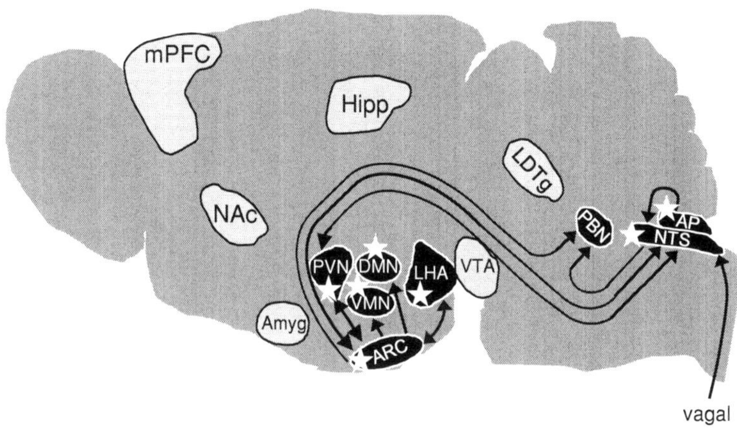

Neuronal Circuits Controlling Hedonic Eating

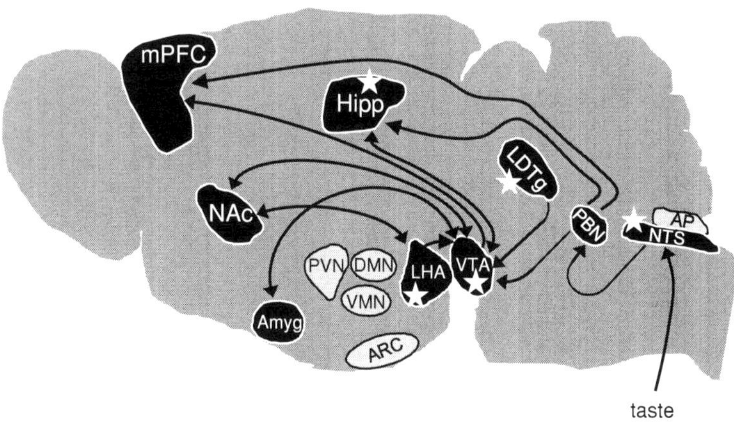

Fig. 1 Model of ghrelin action on neuronal circuits controlling homeostatic and hedonic eating. Cartoons represent sagittal slices of rodent brain depicting brain circuits implicated in ghrelin's regulation of the homeostatic (*upper panel*) or hedonic (*lower panel*) aspects of eating. *Black* areas represent brain nuclei involved in each circuit, and *arrows* indicate probable connections between those brain nuclei. Stars label brain nuclei where GHSR is expressed. Abbreviations: *Amyg* amygdala, *AP* area postrema, *ARC* arcuate nucleus, *DMN* dorsomedial nucleus, *Hipp* hippocampus, *LDTg* laterodorsal tegmental area, *LHA* lateral hypothalamic area, *mPFC* medial prefrontal cortex, *NAc* nucleus accumbens, *NTS* nucleus tractus solitaries, *PBN* parabrachial nucleus, *PVN* paraventricular nucleus of the hypothalamus, *VMN* ventromedial nucleus, *VTA* ventral tegmental area

mechanism by which ghrelin receptor 1a regulates food intake involves its dimerization with other G protein-coupled receptors. The ghrelin receptor 1a has been shown to heterodimerize with the melanocortin 3 receptor, the serotonin 2C receptor, and the dopamine receptors, all involved in food intake and food reward regulation (Schellekens et al. 2013; Kern et al. 2012; Jiang et al. 2006; Rediger et al. 2011). Heterodimerization could serve to modulate specific functions of the ghrelin receptor 1a, such as signaling pathways, or to act as an allosteric mechanism to regulate signaling pathways of the other receptors, independently of ghrelin binding (Schellekens et al. 2013; Kern et al. 2012; Jiang et al. 2006; Rediger et al. 2011).

Modulation of Hedonic Aspects of Eating in Rodent Models by Ghrelin

Evidence from Studies Using Pharmacological Manipulations of the Ghrelin System

Evidence shows that ghrelin enhances preference for pleasurable, sweet, and fatty foods. In this regard, ghrelin administration shifts food preference toward a high-fat diet (HFD) (Shimbara et al. 2004). Ghrelin administration also increases intake of palatable saccharin solution and preference for saccharin-flavored foods in mice (Disse et al. 2010). Similarly, rats treated with a ghrelin receptor 1a antagonist consume less peanut butter and the liquid nutritional supplement Ensure®, but do not change intake of regular chow in a free choice protocol (Egecioglu et al. 2010). Likewise, treatment with a ghrelin receptor 1a antagonist selectively decreases intake of sucrose solution in rats and saccharin solution self-administration in mice (Landgren et al. 2011).

Ghrelin also enhances the motivation to obtain preferred foods, as evaluated by operant lever-pressing or operant nose-poking behavioral tasks in progressive ratio paradigms. Ghrelin administration increases operant lever-pressing for sucrose, peanut butter-flavored sucrose or HFD pellets in rodents (Perello et al. 2010; Finger et al. 2012; Skibicka et al. 2011; Overduin et al. 2012). Conversely, treatment with a ghrelin receptor 1a antagonist reduces operant responding for sucrose solution (Landgren et al. 2011). In addition, ghrelin increases food anticipatory activity, which is characterized by increased arousal, increased locomotor activity, and an elevated body temperature in anticipation of a predicted meal (Merkestein et al. 2012; Jerlhag et al. 2006). Also, ghrelin secreted in anticipation of a meal correlates to anticipatory locomotor activity, and administration of ghrelin increases locomotor activity and foraging-like activities in rodents (Blum et al. 2009; Keen-Rhinehart and Bartness 2005; Jerlhag et al. 2007). On the other hand, ghrelin receptor 1a antagonists decrease anticipatory behavior for a palatable meal (Merkestein et al. 2012).

Ghrelin can also affect more complex, reward-related eating behaviors such as those that take place in a food conditioned place preference (CPP) test. In the food CPP test, animals are conditioned to associate one chamber of the CPP apparatus with regular chow and a second, visually and texturally distinct chamber with an equal-calorie amount of a more pleasurable food, such as HFD. After conditioning, animals have free access to both chambers in the absence of food, and conditioned place preference for HFD is demonstrated by animals spending more time in the chamber associated with the more rewarding food. Food CPP studies performed in mice reveal that both administration of ghrelin and physiological increases in plasma ghrelin induced by caloric restriction enable acquisition of CPP for HFD (Perello et al. 2010; Disse et al. 2011). Similarly, treatment with a ghrelin receptor 1a antagonist blocks CPP for chocolate pellets in satiated rats (Egecioglu et al. 2010). Of note, the assessment of the ghrelin effect on the hedonic valuation *per se* by monitoring the avidity of ingestion of a liquid food via lickometry has suggested that ghrelin does not affect food palatability (Overduin et al. 2012).

The dopaminergic VTA neurons are important for ghrelin's effects on hedonic aspects of eating. Exogenous ghrelin releases dopamine in the NAc from VTA neuronal terminals, and ghrelin increases action potential frequency in dopaminergic VTA neurons (Abizaid et al. 2006; McCallum et al. 2011; Jerlhag 2008; Jerlhag et al. 2006, 2007). Acute intra-VTA administration of ghrelin increases intake of regular food, intake of peanut butter over regular chow, and operant lever-pressing for sucrose and banana-flavored pellets (Abizaid et al. 2006; Naleid et al. 2005; Egecioglu et al. 2010; Skibicka et al. 2011; Weinberg et al. 2011). In addition, pretreatment with a dopamine D1 receptor antagonist eliminates ghrelin-induced increases in lever pressing in rats, without compromising generalized motor control, indicating a role for dopamine signaling in ghrelin's motivational feeding effects (Overduin et al. 2012). On the other hand, intra-VTA administration of ghrelin receptor 1a antagonists decreases food intake in response to peripherally administrated ghrelin, intake of a more preferred HFD, and fasting-induced operant lever pressing for sucrose pellets (Abizaid et al. 2006; Naleid et al. 2005; King et al. 2011; Skibicka et al. 2011). Chronic intra-VTA administration of ghrelin also dose-dependently increases intake of regular chow (King et al. 2011), and VTA-lesioned rats spend less time than control rats exploring tubes containing peanut butter in response to centrally administrated ghrelin (Egecioglu et al. 2010). Similar effects are observed in food-restricted rats, in which chronic intra-VTA administration of ghrelin enhances while chronic intra-VTA delivery of a ghrelin receptor 1a antagonist blunts operant responding for chocolate-flavored pellets (King et al. 2011). Furthermore, intra-VTA administration of ghrelin fails to affect operant lever-pressing for food rewards in animals with dopamine depletion induced by delivery of the neurotoxin 6-hydroxydopamine in the VTA (Weinberg et al. 2011). Ghrelin administration into the VTA also stimulates locomotor activity via an increase in the extracellular concentration of dopamine in the NAc (Jerlhag et al. 2007).

The rest of the neuronal circuit recruited by ghrelin to regulate hedonic aspects of eating is just starting to be elucidated. Ghrelin action on food reward requires intact orexin signaling, as evidenced by the failure of orexin-knockout mice or wild-type (WT) mice given an orexin receptor antagonist to manifest ghrelin-induced effects on HFD reward (Perello et al. 2010). Other signals that likely mediate ghrelin actions on food intake are the endocannabinoids, which regulate both homeostatic and hedonic aspects of eating (Harrold and Williams 2003). Central injection of ghrelin to endocannabinoid receptor type 1 knockout mice fails to increase food intake, suggesting that the endocannabinoid signaling is necessary for ghrelin's orexigenic effect (Kola et al. 2008). Moreover, the ghrelin-induced enhancement of food CPP seems to be partially mediated by the cholinergic pathway (Disse et al. 2011). In this regard, nicotinic receptor signaling seems to play a role in ghrelin's actions on food reward since administration of a selective antagonist of the $\alpha3\beta4$ nicotinic receptor blocks both ghrelin-induced increase of sucrose intake and dopamine release in the NAc following intra-VTA administration of ghrelin (McCallum et al. 2011). The stimulatory effect of ghrelin on dopaminergic neurons of the VTA also appears to depend on the excitatory glutamatergic inputs (Abizaid et al. 2006). In fact, the ability of ghrelin to activate the dopaminergic VTA system and the locomotor activity is suppressed by pharmacological blockade of glutamatergic N-methyl-D-aspartate (NMDA) receptors but not by blockade of opioid or orexin receptors (Jerlhag et al. 2011).

Evidence from Studies Using Genetic Manipulations of the Ghrelin System

Mouse models with genetic manipulations of the ghrelin system have been instrumental in order to establish the mechanisms underlying ghrelin's actions on eating behaviors. These models include mice over-expressing ghrelin and mice with deletion of the genes encoding ghrelin, ghrelin receptor 1a, or the enzyme that octanoylates ghrelin [ghrelin O-acyltransferase (GOAT)]. In addition, a conditional ghrelin receptor 1a null mouse model in which ghrelin receptor 1a transcription is globally blocked but can be cell-specifically reactivated in a Cre recombinase-mediated fashion has been generated.

Most mouse models overexpressing or lacking bioactive ghrelin show minor alterations on food intake behaviors. Transgenic mice with increased brain and circulating bioactive ghrelin do not differ from WT controls in food intake or body weight (Reed et al. 2008). In contrast, chronic overproduction of bioactive ghrelin in the stomach increases food intake but does not alter long-term body weight gain due to a paradoxical increase in energy expenditure (Bewick et al. 2009). The double-transgenic mice overexpressing both human ghrelin and GOAT genes in the liver have decreased energy expenditure and increased body weight without food intake alterations only when fed on HFD rich in medium-chain triglycerides (Kirchner et al. 2009). Similarly, ghrelin-deficient mice show normal food intake and body weight, as compared to WT mice. (De Smet et al. 2006; Wortley et al. 2005; Sun et al. 2003; Dezaki et al. 2006; Sato et al. 2008). In addition, no differences are observed when some other aspects of eating behaviors of ghrelin-deficient mice are evaluated, including post-fasting hyperphagia or forced dark cycle induced eating (Wortley et al. 2005; Sun et al. 2003; Pfluger et al. 2008; Sato et al. 2008; De Smet et al. 2006). Of note, ghrelin-deficient mice show some alterations in their food intake behaviors under particular experimental settings. For instance, they lack anticipatory eating response failing to match the increase in food intake observed in WT type controls during 6 h food intake following repeated overnight fasts (Abizaid et al. 2006). Studies where ghrelin-deficient mice were chronically fed with HFD failed to show any reduction of food intake (Dezaki et al. 2006; Wortley et al. 2005; Sun et al. 2003). Only one of these studies was able to detect that ghrelin deficiency results in reduced body weight and fat mass, among other beneficial effects (Wortley et al. 2005). On the other hand, the GOAT-deficient mice, which lack plasma bioactive ghrelin, do not differ from WT controls in food intake or body weight, when fed with regular chow (Kirchner et al. 2009; Zhao et al. 2010). One study showed that GOAT deficiency results in decreased body weight when animals were fed on HFD rich in medium-chain triglycerides (Kirchner et al. 2009), but this body weight phenotype was not observed by other researchers (Zhao et al. 2010). GOAT-deficient mice display an attenuated motivation for HFD in an operant responding model and also a decreased hedonic feeding response examined in a "dessert effect" protocol, in which the intake of a palatable HFD pellet "dessert" is assessed in calorically sated mice (Davis et al. 2012).

The use of ghrelin receptor 1a deficient mice has shown an obligatory role of ghrelin signaling in certain hedonic aspects of eating that are separated from eating associated with body weight homeostasis. Ghrelin receptor 1a deficient mice show a subtle but significant decrease in body weight without food intake alterations when they have free access to regular chow diet (Abizaid et al. 2006; Zigman et al. 2005; Sun et al. 2004). Interestingly, ghrelin receptor 1a null mice are resistant to HFD-induced body weight gain, if they are exposed to HFD early in their life (Zigman et al. 2005; Perello et al. 2012). However, no differences in HFD-induced body weight gain are observed if mice are exposed to HFD during adulthood (Sun et al. 2008). Additionally, ghrelin receptor 1a deficient mice show an improvement of aging-associated obesity due mainly to a reduced adiposity and increased thermogenesis (Lin et al. 2011; Ma et al. 2011). Ghrelin/ghrelin receptor 1a double knockout mice exhibit decreased body weight when placed on a standard chow diet (Pfluger et al. 2008). Ghrelin receptor 1a deficient mice are protected from the weight gain induced by exposure to HFD although no reduction in HFD intake is observed (Zigman et al. 2005; Perello et al. 2012). Importantly, ghrelin receptor 1a deficient mice have a reduced intake of the more rewarding food in a free choice paradigm and a reduced dopamine release in the NAc induced by rewarding foods (Egecioglu et al. 2010). Also, ghrelin receptor 1a null mice also fail to enhance feeding in response to a light cue used as positive-conditioned stimulus as compared to WT mice (Walker et al. 2012).

The significance of ghrelin signaling on hedonic eating regulation becomes more evident in situations in which plasma ghrelin is physiologically elevated, such as fasting, caloric restriction, or stress (Perello and Zigman 2012). In this regard, ghrelin receptor 1a deficient mice show important eating behavior alterations under specific experimental conditions. For instance, WT mice subjected to prolonged caloric restriction show enhanced-CPP for HFD while ghrelin receptor

1a deficient mice lack such response (Perello et al. 2010; Disse et al. 2011). Moreover, ghrelin receptor 1a deficient mice in response to scheduled meals have both attenuated anticipatory hyperlocomotion and reduced expression of the marker of cellular activation c-fos in the mesolimbic pathway (Lamont et al. 2012; Blum et al. 2009). Similarly, ghrelin receptor 1a deficient mice do not anticipate food when exposed to an activity-based anorexia model, in which mice are given free access to a running wheel and fed once per day for 2 h (Verhagen et al. 2011). The chronic social defeat stress (CSDS) procedure, which subjects mice to daily bouts of social defeat by aggressive male mice, has been also used to study the physiological effect of ghrelin on feeding behaviors (Lutter et al. 2008; Patterson et al. 2013). WT mice exposed to CSDS increase their plasma ghrelin concentration and regular chow intake during and for at least 1 month after the defeat period. In contrast, ghrelin receptor 1a null mice fail to show CSDS-induced hyperphagia (Lutter et al. 2008; Patterson et al. 2013). In WT mice, CSDS also increases CPP for HFD while such a stress-induced food reward response is not observed in CSDS-exposed ghrelin receptor 1a null mice (Chuang et al. 2011). In contrast to these findings, a chronic unpredictable stress model that also elevates plasma ghrelin decreases food intake and body weight gain in WT mice, while similarly treated ghrelin receptor 1a deficient mice lack these changes (Patterson et al. 2010). Thus, further work is needed to clarify the role of ghrelin on food intake among different rodent models of stress.

The mouse model with reactivable genetic deletion of ghrelin receptor 1a has been very valuable to establish the physiological roles of some of ghrelin's brain targets. In this nontraditional mouse model, ghrelin receptor 1a gene expression is disrupted by a transcriptional blocking cassette flanked by loxP sites that enable Cre recombinase-mediated ghrelin receptor 1a gene re-expression (Zigman et al. 2005). Thus, the ghrelin receptor 1a transcription is globally blocked in ghrelin receptor 1a null mice, but it can be cell-specifically reactivated in a Cre-mediated fashion (Zigman et al. 2005). Using this strategy, mice expressing ghrelin receptor 1a selectively in tyrosine hydroxylase-containing cells, including a subset of VTA dopaminergic neurons, was generated (Chuang et al. 2011). These mice show a significant, albeit reduced, response to the orexigenic effects of ghrelin (Chuang et al. 2011). Interestingly, mice with re-expression of ghrelin receptor 1a selectively in tyrosine hydroxylase-containing neurons show full CPP for HFD when treated with exogenous ghrelin or exposed to a CSDS protocol (Chuang et al. 2011). This study suggests that expression of ghrelin receptor 1a in dopaminergic neurons is sufficient for ghrelin's actions on both food intake and food reward. Of note, mice with re-expression of ghrelin receptor 1a in specific hindbrain nuclei, including the NTS, DMV, AP, nucleus ambiguous, and facial motor nucleus, fail to show ghrelin-induced food intake (Scott et al. 2012). Thus, direct action of circulating ghrelin on ghrelin receptor 1a expressing hindbrain neurons is not sufficient to mediate acute orexigenic effects of ghrelin.

Relevance of Ghrelin Effects on Hedonic Aspects of Eating for Humans

Many studies suggest that ghrelin signaling is relevant for human food intake regulation. Human beings have a preprandial rise and a postprandial decline in plasma ghrelin levels suggesting that ghrelin recapitulates in humans its physiological role in hunger and/or meal initiation observed in rodents (Cummings 2006; Cummings et al. 2001). The preprandial ghrelin surge occurs as many times per day as meals are provided to subjects exposed to habituated feeding schedules (Cummings 2006; Cummings et al. 2001). Importantly, ghrelin levels also rise preprandially initiating meals voluntarily in the absence of cues related to time or food, and the temporal profiles of plasma ghrelin levels and hunger scores tightly overlap in this setting (Cummings 2006; Cummings et al. 2001). The postprandial ghrelin decrease seems to be critical for satiety sensation and, accordingly, it decreases proportionally to meal calorie content (le Roux et al. 2005). Of note, postprandial ghrelin decrease is impaired after high-fat meals likely contributing to reduce satiety and causing overeating (Yang et al. 2009). The mechanisms involved in the control of pre and postprandial ghrelin regulation in humans are currently unclear.

Most studies show that intravenous bolus or continuous administration of ghrelin stimulates hunger sensations and food intake in healthy individuals (Akamizu et al. 2008; Adachi et al. 2010; Schmid et al. 2005; Levin et al. 2006; Wren et al. 2001; Falken et al. 2010; Druce et al. 2005). It is interesting to note that some of these studies have used ghrelin doses that result in supra-physiological increases in plasma hormone levels. Also, administration of exogenous ghrelin cannot mimic the postprandial decrease of the hormone levels that occur in physiological conditions. Despite these considerations, it is normally accepted that exogenous ghrelin can regulate meal initiation and food intake of human beings (Cummings 2006). Functional magnetic resonance imaging studies indicate that ghrelin increases the neural response in brain centers implicated in hedonic feeding of human subjects (Goldstone et al. 2009; Malik et al. 2008; Neary and Batterham 2010). Fasting-induced increases of plasma ghrelin enhance both the appeal of high-calorie more than low-calorie foods and the reward-related brain centers' response to pictures of high-calorie over low-calorie foods (Goldstone et al. 2009). Also, ghrelin administration to human subjects increases the activation of some hedonic feeding-related brain centers, including the substance nigra and the VTA, in response to tempting food pictures (Malik et al. 2008; Neary and Batterham 2010). Thus, ghrelin seems to have a significant role in food reward behavior and appetite regulation in humans.

Role of Ghrelin and Ghrelin Receptor 1a on Disorders with Alterations of Food Intake

Obesity. Obesity is defined as an excessive fat accumulation that presents a risk to health. Obesity is a heterogeneous disorder with several potential etiologies including genetic and environmental factors. Little association has been found between obesity and ghrelin or ghrelin receptor 1a mutations in humans (Gueorguiev et al. 2009; Liu et al. 2011). However, the ghrelin system appears relevant for human obesity (Hillman et al. 2011). Most obese patients have chronically low levels of circulating ghrelin and a blunting of the nocturnal plasma ghrelin increase compared to normal subjects (Hillman et al. 2011; Tschop et al. 2001). Similarly, plasma ghrelin is decreased in diet-induced obesity mouse models, where a resistance to ghrelin-induced food intake and ghrelin-induced motivation to obtain food rewards is observed (Finger et al. 2012; Perreault et al. 2004; Briggs et al. 2010). Still, obese people seem to be fully sensitive to the orexigenic effects of exogenous ghrelin (Druce et al. 2005). Several studies show that obese people have a blunted postprandial decrease of plasma ghrelin, which likely increases the time they feel hungry and participates in the pathophysiology of obesity (le Roux et al. 2005; Yang et al. 2009; Morpurgo et al. 2003; English et al. 2002). Also, ghrelin levels rise in obese individuals after weight loss induced by dieting, and such increase of plasma ghrelin likely contributes to the rebound weight gain commonly observed in dieters (Cummings et al. 2002b). In addition, the marked and prolonged weight loss observed in obese individuals who undergo Roux-en-Y gastric bypass surgery is thought to be enhanced by postsurgery reductions in circulating ghrelin (Cummings and Shannon 2003; Beckman et al. 2010). These clinical studies, among others (Schellekens et al. 2012), support the concept that pharmacological manipulations of ghrelin signaling may be a potential strategy to reduce food intake and ultimately body weight in obese patients (See "Ghrelin Receptors a Novel Target for Obesity" for details).

Prader-Willi syndrome (PWS). PWS is a genetic obesity syndrome caused by a defect in the chromosome 15 (q11–13). Children with PWS display growth hormone deficiency, rapid weight gain, and voracious appetite. Hyperphagia of PWS seems to involve alterations of hedonic aspects of feeding, since functional magnetic resonance imaging in these patients shows enhanced activation of the mesolimbic system areas following regular meals intake, when high-calorie foods are offered or even when food pictures are displayed to them (Miller et al. 2007; Holsen et al. 2006; Dimitropoulos and Schultz 2008). Of note, most PWS patients have several-fold higher ghrelin levels compared to weight-matched controls (Cummings et al. 2002a; DelParigi et al. 2002; Haqq et al. 2003a). In some PWS patients, the hyperphagia is related to high plasma ghrelin as hyperghrelinemia precedes obesity and plasma ghrelin levels positively correlate with their feelings of hunger (Haqq et al. 2003a; Purtell et al. 2011; Feigerlova et al. 2008). Of note, not all young PWS patients have elevated plasma ghrelin levels (Haqq et al. 2008). In addition, intervention studies suppressing ghrelin levels in PWS patients have failed to reduce appetite or compulsive eating (Tan et al. 2004; De Waele et al. 2008; Haqq et al. 2003b). Thus, the role of the ghrelin system in the pathogenesis of this disorder is still unclear.

Anorexia Nervosa. Anorexia nervosa is an eating disorder of unknown etiology characterized by refusal to maintain a minimally required healthy weight, intense fear of gaining weight, and misinterpretation of body shape. Anorexia nervosa can be divided into a restrictive type, with reduced food intake, and a binge eating/ purging type, with binge eating/purging episodes during anorexia phases. Most studies report that fasted anorexia nervosa patients show high ghrelin levels, which normalize after food intake or body weight recovery (Ogiso et al. 2011). Patients with binging/purging anorexia nervosa type have higher ghrelin levels (Tanaka et al. 2003, 2004). Also, single nucleotide polymorphisms in ghrelin gene are specifically associated with binging/purging anorexia nervosa type (Dardennes et al. 2007). However, these findings have not been fully reproduced by other studies (Cardona Cano et al. 2012). Thus, the pathophysiological implications of high plasma ghrelin in anorexia nervosa are currently unclear. It has been proposed that administration of ghrelin (or ghrelin agonists) could increase food intake and hunger in these patients and thus promote weight gain. Until now, three studies have evaluated the effect of ghrelin administration on anorexia nervosa patients (Miljic et al. 2006; Broglio et al. 2004; Hotta et al. 2009). In one study, anorexia nervosa patients felt significantly less hungry compared to the thin control subjects, suggesting that anorexia nervosa patients are resistant to the orexigenic effects of ghrelin (Miljic et al. 2006). However, other studies found increased hunger sensation and increased food intake after ghrelin administration in some patients with anorexia nervosa (Broglio et al. 2004; Hotta et al. 2009). Thus, further studies are needed to determine if ghrelin treatment is a therapeutic option for this disorder.

Bulimia nervosa. Bulimia nervosa is a psychiatric disorder characterized by repetitive episodes of consumption of large amounts of food followed by compensatory behaviors in order to prevent weight gain, including self-induced vomiting, laxative abuse, and excessive exercising. As discussed in a recent review, findings from many studies that have investigated the potential pathophysiological role of ghrelin in the bulimia nervosa are inconsistent, and it is currently unclear whether the ghrelin system dysfunctions are relevant in this eating disorder (Cardona Cano et al. 2012).

Binge eating disorders. In contrast to bulimia nervosa, patients who suffer binge eating disorders engage in bouts of binge eating with no compensatory behavior afterwards that increases the risk for obesity. Some patients with binge eating disorders have an altered ghrelin dynamics, characterized by less postprandial decrease of ghrelin with a longer time to nadir compared with obese subjects, that could contribute to larger meals as seen during binge episodes (Geliebter et al. 2005, 2008). However, other studies have shown that fasting plasma ghrelin levels do not correlate with the frequency and severity of binging (Monteleone et al. 2005). Interestingly, a single nucleotide polymorphism of the ghrelin gene has been associated with binge eating disorders (Monteleone et al. 2007). As for other eating disorders, further studies are necessary to establish a link between binge eating disorders and ghrelin.

Cachexia. Cachexia or wasting syndrome is defined as unintentional appetite and body mass loss that cannot be reversed nutritionally. Lean body mass is lost even when the affected patient eats more calories, indicating that body mass loss is due to another primary pathology taking place. Cachexia is seen in patients with cancer, acquired immunodeficiency syndrome, chronic obstructive pulmonary disease, chronic renal insufficiency, congestive heart failure, tuberculosis, among others. Total plasma ghrelin levels are elevated in patients with cachexia, as expected for a chronic state of energy deficiency (DeBoer 2008). Despite the elevated plasma ghrelin concentrations, patients with cachexia remain sensitive to the orexigenic effects of ghrelin. Clinical studies have shown that administration of ghrelin or ghrelin receptor 1a agonists increased both food intake and body weight in patients with cachexia secondary to congestive heart failure, chronic obstructive pulmonary disease, or chronic renal insufficiency (Nagaya et al. 2004, 2005; Wynne et al. 2005; Deboer et al. 2008; Ashby et al. 2009). In addition, several trials have demonstrated the efficacy and safety of ghrelin or ghrelin receptor 1a agonists to increase food intake and body weight in patients with cancer-associated cachexia (Neary et al. 2004; Strasser et al. 2008; Garcia et al. 2013). Thus, ghrelin system may be a potential pharmacological target in the treatment of cachexia (Argiles and Stemmler 2013).

Concluding Remarks

Recent studies have started to reveal the complex neuronal circuits and mechanisms by which ghrelin promotes food intake. Ghrelin not only acts on neuronal circuits that regulate homeostatic intake of food but also on neuronal circuits that affect hedonic aspects of eating including preference for palatable foods, motivation to obtain preferred foods, food anticipatory locomotor activity, rewarding value of preferred foods, and acquisition of food CPP. Thus, ghrelin modulates a variety of key aspects of hedonic eating that directly impact on feeding behaviors. Of note, other peripheral signals from adipose tissue (e.g., leptin), pancreas (e.g., insulin), and the gastrointestinal tract (e.g., peptide YY, glucagon-like peptide-1, cholecystokinin) also regulate central circuits controlling food intake. However, ghrelin is the only known peptide hormone that causes an acute and potent increase of food intake when administrated in small doses to animals or human beings. This unique feature makes the ghrelin system exceptionally attractive for the development of specific pharmacological therapies to treat eating disorders.

Acknowledgments This manuscript was supported by grants from the National Agency of Scientific and Technological Promotion of Argentina (PICT2010-1954 and PICT2011-2142 to MP, and PICT2010-1589 and PICT2011-1816 to JR). We would like to thank Nicolas De Francesco and Agustina Cabral for critically reading the manuscript.

References

Abizaid A, Liu ZW, Andrews ZB, Shanabrough M, Borok E, Elsworth JD, Roth RH, Sleeman MW, Picciotto MR, Tschop MH, Gao XB, Horvath TL (2006) Ghrelin modulates the activity and synaptic input organization of midbrain dopamine neurons while promoting appetite. J Clin Invest 116(12):3229–3239. doi:10.1172/JCI29867

Adachi S, Takiguchi S, Okada K, Yamamoto K, Yamasaki M, Miyata H, Nakajima K, Fujiwara Y, Hosoda H, Kangawa K, Mori M, Doki Y (2010) Effects of ghrelin administration after total gastrectomy: a prospective, randomized, placebo-controlled phase II study. Gastroenterology 138(4):1312–1320. doi:10.1053/j.gastro.2009.12.058, S0016-5085(10)00010-7 [pii]

Akamizu T, Iwakura H, Ariyasu H, Hosoda H, Murayama T, Yokode M, Teramukai S, Seno H, Chiba T, Noma S, Nakai Y, Fukunaga M, Kangawa K (2008) Repeated administration of ghrelin to patients with functional dyspepsia: its effects on food intake and appetite. Eur J Endocrinol 158(4):491–498. doi:10.1530/EJE-07-0768, 158/4/491 [pii]

Argiles JM, Stemmler B (2013) The potential of ghrelin in the treatment of cancer cachexia. Expert Opin Biol Ther 13(1):67–76. doi:10.1517/14712598.2013.727390

Ashby DR, Ford HE, Wynne KJ, Wren AM, Murphy KG, Busbridge M, Brown EA, Taube DH, Ghatei MA, Tam FW, Bloom SR, Choi P (2009) Sustained appetite improvement in malnourished dialysis patients by daily ghrelin treatment. Kidney Int 76(2):199–206. doi:10.1038/ki.2009.114, ki2009114 [pii]

Atasoy D, Betley JN, Su HH, Sternson SM (2012) Deconstruction of a neural circuit for hunger. Nature 488(7410):172–177. doi:10.1038/nature11270, nature11270 [pii]

Banks WA (2008) The blood-brain barrier: connecting the gut and the brain. Regul Pept 149(1–3):11–14. doi:10.1016/j.regpep.2007.08.027, S0167-0115(08)00061-X [pii]

Beckman LM, Beckman TR, Earthman CP (2010) Changes in gastrointestinal hormones and leptin after Roux-en-Y gastric bypass procedure: a review. J Am Diet Assoc 110(4):571–584. doi:10.1016/j.jada.2009.12.023, S0002-8223(09)02093-8 [pii]

Berthoud HR (2011) Metabolic and hedonic drives in the neural control of appetite: who is the boss? Curr Opin Neurobiol 21(6):888–896. doi:10.1016/j.conb.2011.09.004, S0959-4388(11)00148-6[pii]

Bewick GA, Kent A, Campbell D, Patterson M, Ghatei MA, Bloom SR, Gardiner JV (2009) Mice with hyperghrelinemia are hyperphagic and glucose intolerant and have reduced leptin sensitivity. Diabetes 58(4):840–846. doi:10.2337/db08-1428, db08-1428 [pii]

Blum ID, Patterson Z, Khazall R, Lamont EW, Sleeman MW, Horvath TL, Abizaid A (2009) Reduced anticipatory locomotor responses to scheduled meals in ghrelin receptor deficient mice. Neuroscience 164(2):351–359. doi:10.1016/j.neuroscience.2009.08.009, S0306-4522(09)01292-5 [pii]

Briggs DI, Andrews ZB (2011) Metabolic status regulates ghrelin function on energy homeostasis. Neuroendocrinology 93(1):48–57. doi:10.1159/000322589, 000322589 [pii]

Briggs DI, Enriori PJ, Lemus MB, Cowley MA, Andrews ZB (2010) Diet-induced obesity causes ghrelin resistance in arcuate NPY/AgRP neurons. Endocrinology 151(10):4745–4755. doi:10.1210/en.2010-0556, en.2010-0556, [pii]

Broglio F, Gianotti L, Destefanis S, Fassino S, Abbate Daga G, Mondelli V, Lanfranco F, Gottero C, Gauna C, Hofland L, Van der Lely AJ, Ghigo E (2004) The endocrine response to acute ghrelin administration is blunted in patients with anorexia nervosa, a ghrelin hypersecretory state. Clin Endocrinol (Oxf) 60(5):592–599. doi:10.1111/j.1365-2265.2004.02011.x, CEN2011 [pii]

Cardona Cano S, Merkestein M, Skibicka KP, Dickson SL, Adan RA (2012) Role of ghrelin in the pathophysiology of eating disorders: implications for pharmacotherapy. CNS Drugs 26(4):281–296. doi:10.2165/11599890-000000000-00000, 1 [pii]

Chuang JC, Perello M, Sakata I, Osborne-Lawrence S, Savitt JM, Lutter M, Zigman JM (2011) Ghrelin mediates stress-induced food-reward behavior in mice. J Clin Invest 121(7):2684–2692. doi:10.1172/JCI57660, 57660 [pii]

Cong WN, Golden E, Pantaleo N, White CM, Maudsley S, Martin B (2010) Ghrelin receptor signaling: a promising therapeutic target for metabolic syndrome and cognitive dysfunction. CNS Neurol Disord Drug Targets 9(5):557–563. doi:BSP/CDTCNSND/E-Pub/00055, [pii]

Cowley MA, Smith RG, Diano S, Tschop M, Pronchuk N, Grove KL, Strasburger CJ, Bidlingmaier M, Esterman M, Heiman ML, Garcia-Segura LM, Nillni EA, Mendez P, Low MJ, Sotonyi P, Friedman JM, Liu H, Pinto S, Colmers WF, Cone RD, Horvath TL (2003) The distribution and mechanism of action of ghrelin in the CNS demonstrates a novel hypothalamic circuit regulating energy homeostasis. Neuron 37(4):649–661. doi:S0896627303000631, [pii]

Cummings DE (2006) Ghrelin and the short- and long-term regulation of appetite and body weight. Physiol Behav 71:71–84

Cummings DE, Clement K, Purnell JQ, Vaisse C, Foster KE, Frayo RS, Schwartz MW, Basdevant A, Weigle DS (2002a) Elevated plasma ghrelin levels in Prader Willi syndrome. Nat Med 8(7):643–644. doi:10.1038/nm0702-643, nm0702-643 [pii]

Cummings DE, Purnell JQ, Frayo RS, Schmidova K, Wisse BE, Weigle DS (2001) A preprandial rise in plasma ghrelin levels suggests a role in meal initiation in humans. Diabetes 50(8):1714–1719

Cummings DE, Shannon MH (2003) Ghrelin and gastric bypass: is there a hormonal contribution to surgical weight loss? J Clin Endocrinol Metab 88(7):2999–3002

Cummings DE, Weigle DS, Frayo RS, Breen PA, Ma MK, Dellinger EP, Purnell JQ (2002b) Plasma ghrelin levels after diet-induced weight loss or gastric bypass surgery. N Engl J Med 346(21):1623–1630. doi:10.1056/NEJMoa012908, 346/21/1623 [pii]

Damian M, Marie J, Leyris JP, Fehrentz JA, Verdie P, Martinez J, Baneres JL, Mary S (2012) High constitutive activity is an intrinsic feature of ghrelin receptor protein: a study with a functional monomeric GHS-R1a receptor reconstituted in lipid discs. J Biol Chem 287(6):3630–3641. doi:10.1074/jbc.M111.288324, M111.288324 [pii]

Dardennes RM, Zizzari P, Tolle V, Foulon C, Kipman A, Romo L, Iancu-Gontard D, Boni C, Sinet PM, Therese Bluet M, Estour B, Mouren MC, Guelfi JD, Rouillon F, Gorwood P, Epelbaum J (2007) Family trios analysis of common polymorphisms in the obestatin/ghrelin, BDNF and AGRP genes in patients with Anorexia nervosa: association with subtype, body-mass index, severity and age of onset. Psychoneuroendocrinology 32(2):106–113. doi:10.1016/j.psyneuen.2006.11.003, S0306-4530(06)00197-1 [pii]

Date Y (2012) Ghrelin and the vagus nerve. Methods Enzymol 514:261–269. doi:10.1016/B978-0-12-381272-8.00016-7, B978-0-12-381272-8.00016-7 [pii]

Date Y, Murakami N, Toshinai K, Matsukura S, Niijima A, Matsuo H, Kangawa K, Nakazato M (2002) The role of the gastric afferent vagal nerve in ghrelin-induced feeding and growth hormone secretion in rats. Gastroenterology 123(4):1120–1128. doi:S0016508502002172, [pii]

Date Y, Shimbara T, Koda S, Toshinai K, Ida T, Murakami N, Miyazato M, Kokame K, Ishizuka Y, Ishida Y, Kageyama H, Shioda S, Kangawa K, Nakazato M (2006) Peripheral ghrelin transmits orexigenic signals through the noradrenergic pathway from the hindbrain to the hypothalamus. Cell Metab 4(4):323–331. doi:10.1016/j.cmet.2006.09.004, S1550-4131(06)00303-2 [pii]

Davis JF, Perello M, Choi DL, Magrisso IJ, Kirchner H, Pfluger PT, Tschoep M, Zigman JM, Benoit SC (2012) GOAT induced ghrelin acylation regulates hedonic feeding. Horm Behav 62(5):598–604. doi:10.1016/j.yhbeh.2012.08.009, S0018-506X(12)00202-4 [pii]

De Smet B, Depoortere I, Moechars D, Swennen Q, Moreaux B, Cryns K, Tack J, Buyse J, Coulie B, Peeters TL (2006) Energy homeostasis and gastric emptying in ghrelin knockout mice. J Pharmacol Exp Ther 316(1):431–439. doi:10.1124/jpet.105.091504, jpet.105.091504 [pii]

De Waele K, Ishkanian SL, Bogarin R, Miranda CA, Ghatei MA, Bloom SR, Pacaud D, Chanoine JP (2008) Long-acting octreotide treatment causes a sustained decrease in ghrelin concentrations but does not affect weight, behaviour and appetite in subjects with Prader-Willi syndrome. Eur J Endocrinol 159(4):381–388. doi:10.1530/EJE-08-0462, EJE-08-0462 [pii]

DeBoer MD (2008) Emergence of ghrelin as a treatment for cachexia syndromes. Nutrition 24(9):806–814. doi:10.1016/j.nut.2008.06.013, S0899-9007(08)00291-8 [pii]

Deboer MD, Zhu X, Levasseur PR, Inui A, Hu Z, Han G, Mitch WE, Taylor JE, Halem HA, Dong JZ, Datta R, Culler MD, Marks DL (2008) Ghrelin treatment of chronic kidney disease: improvements in lean body mass and cytokine profile. Endocrinology 149(2):827–835. doi:10.1210/en.2007-1046, en.2007-1046 [pii]

DelParigi A, Tschop M, Heiman ML, Salbe AD, Vozarova B, Sell SM, Bunt JC, Tataranni PA (2002) High circulating ghrelin: a potential cause for hyperphagia and obesity in prader-willi syndrome. J Clin Endocrinol Metab 87(12):5461–5464

Dezaki K, Sone H, Koizumi M, Nakata M, Kakei M, Nagai H, Hosoda H, Kangawa K, Yada T (2006) Blockade of pancreatic islet-derived ghrelin enhances insulin secretion to prevent high-fat diet-induced glucose intolerance. Diabetes 55(12):3486–3493. doi:10.2337/db06-0878, 55/12/3486 [pii]

Diano S, Farr SA, Benoit SC, McNay EC, da Silva I, Horvath B, Gaskin FS, Nonaka N, Jaeger LB, Banks WA, Morley JE, Pinto S, Sherwin RS, Xu L, Yamada KA, Sleeman MW, Tschop MH, Horvath TL (2006) Ghrelin controls hippocampal spine synapse density and memory performance. Nat Neurosci 9(3):381–388. doi:10.1038/nn1656, nn1656 [pii]

Dickson SL, Hrabovszky E, Hansson C, Jerlhag E, Alvarez-Crespo M, Skibicka KP, Molnar CS, Liposits Z, Engel JA, Egecioglu E (2010) Blockade of central nicotine acetylcholine receptor signaling attenuate ghrelin-induced food intake in rodents. Neuroscience 171(4):1180–1186. doi:10.1016/j.neuroscience.2010.10.005, S0306-4522(10)01335-7 [pii]

DiLeone RJ, Taylor JR, Picciotto MR (2012) The drive to eat: comparisons and distinctions between mechanisms of food reward and drug addiction. Nat Neurosci 15(10):1330–1335. doi:10.1038/nn.3202, nn.3202 [pii]

Dimitropoulos A, Schultz RT (2008) Food-related neural circuitry in Prader-Willi syndrome: response to high- versus low-calorie foods. J Autism Dev Disord 38(9):1642–1653. doi:10.1007/s10803-008-0546-x

Disse E, Bussier AL, Deblon N, Pfluger PT, Tschop MH, Laville M, Rohner-Jeanrenaud F (2011) Systemic ghrelin and reward: effect of cholinergic blockade. Physiol Behav 102(5):481–484. doi:10.1016/j.physbeh.2010.12.006, S0031-9384(10)00453-1 [pii]

Disse E, Bussier AL, Veyrat-Durebex C, Deblon N, Pfluger PT, Tschop MH, Laville M, Rohner-Jeanrenaud F (2010) Peripheral ghrelin enhances sweet taste food consumption and preference, regardless of its caloric content. Physiol Behav 101(2):277–281. doi:10.1016/j.physbeh.2010.05.017, S0031-9384(10)00234-9 [pii]

Druce MR, Wren AM, Park AJ, Milton JE, Patterson M, Frost G, Ghatei MA, Small C, Bloom SR (2005) Ghrelin increases food intake in obese as well as lean subjects. Int J Obes (Lond) 29(9):1130–1136. doi:10.1038/sj.ijo.0803001, 0803001 [pii]

Egecioglu E, Jerlhag E, Salome N, Skibicka KP, Haage D, Bohlooly YM, Andersson D, Bjursell M, Perrissoud D, Engel JA, Dickson SL (2010) Ghrelin increases intake of rewarding food in rodents. Addict Biol 15(3):304–311. doi:10.1111/j.1369-1600.2010.00216.x, ADB216 [pii]

English PJ, Ghatei MA, Malik IA, Bloom SR, Wilding JP (2002) Food fails to suppress ghrelin levels in obese humans. J Clin Endocrinol Metab 87(6):2984

Falken Y, Hellstrom PM, Sanger GJ, Dewit O, Dukes G, Gryback P, Holst JJ, Naslund E (2010) Actions of prolonged ghrelin infusion on gastrointestinal transit and glucose homeostasis in humans. Neurogastroenterol Motil 22(6):e192–200. doi:10.1111/j.1365-2982.2009.01463.x, NMO1463 [pii]

Feigerlova E, Diene G, Conte-Auriol F, Molinas C, Gennero I, Salles JP, Arnaud C, Tauber M (2008) Hyperghrelinemia precedes obesity in Prader-Willi syndrome. J Clin Endocrinol Metab 93(7):2800–2805. doi:10.1210/jc.2007-2138, jc.2007-2138 [pii]

Finger BC, Dinan TG, Cryan JF (2012) Diet-induced obesity blunts the behavioural effects of ghrelin: studies in a mouse-progressive ratio task. Psychopharmacology 220(1):173–181. doi:10.1007/s00213-011-2468-0

Fry M, Ferguson AV (2007) The sensory circumventricular organs: brain targets for circulating signals controlling ingestive behavior. Physiol Behav 91(4):413–423. doi:10.1016/j.physbeh.2007.04.003, S0031-9384(07)00130-8 [pii]

Fry M, Ferguson AV (2010) Ghrelin: central nervous system sites of action in regulation of energy balance. Int J Pept 2010. doi:10.1155/2010/616757, 616757 [pii]

Furness JB, Hunne B, Matsuda N, Yin L, Russo D, Kato I, Fujimiya M, Patterson M, McLeod J, Andrews ZB, Bron R (2011) Investigation of the presence of ghrelin in the central nervous system of the rat and mouse. Neuroscience 193:1–9. doi:10.1016/j.neuroscience.2011.07.063, S0306-4522(11)00889-X [pii]

Garcia JM, Friend J, Allen S (2013) Therapeutic potential of anamorelin, a novel, oral ghrelin mimetic, in patients with cancer-related cachexia: a multicenter, randomized, double-blind, crossover, pilot study. Support Care Cancer 21(1):129–137. doi:10.1007/s00520-012-1500-1

Geliebter A, Gluck ME, Hashim SA (2005) Plasma ghrelin concentrations are lower in binge-eating disorder. J Nutr 135(5):1326–1330. doi:135/5/1326, [pii]

Geliebter A, Hashim SA, Gluck ME (2008) Appetite-related gut peptides, ghrelin, PYY, and GLP-1 in obese women with and without binge eating disorder (BED). Physiol Behav 94(5):696–699. doi:10.1016/j.physbeh.2008.04.013, S0031-9384(08)00114-5 [pii]

Gilg S, Lutz TA (2006) The orexigenic effect of peripheral ghrelin differs between rats of different age and with different baseline food intake, and it may in part be mediated by the area postrema. Physiol Behav 87(2):353–359. doi:10.1016/j.physbeh.2005.10.015, S0031-9384(05)00492-0 [pii]

Goldstone AP, Prechtl de Hernandez CG, Beaver JD, Muhammed K, Croese C, Bell G, Durighel G, Hughes E, Waldman AD, Frost G, Bell JD (2009) Fasting biases brain reward systems towards high-calorie foods. Eur J Neurosci 30(8):1625–1635. doi:10.1111/j.1460-9568.2009.06949.x, EJN6949 [pii]

Guan XM, Yu H, Palyha OC, McKee KK, Feighner SD, Sirinathsinghji DJ, Smith RG, Van der Ploeg LH, Howard AD (1997) Distribution of mRNA encoding the growth hormone secretagogue receptor in brain and peripheral tissues. Brain Res Mol Brain Res 48(1):23–29. doi:S0169328X97000715, [pii]

Gueorguiev M, Lecoeur C, Meyre D, Benzinou M, Mein CA, Hinney A, Vatin V, Weill J, Heude B, Hebebrand J, Grossman AB, Korbonits M, Froguel P (2009) Association studies on ghrelin and ghrelin receptor gene polymorphisms with obesity. Obesity (Silver Spring) 17(4):745–754. doi:10.1038/oby.2008.589, oby2008589 [pii]

Haqq AM, Farooqi IS, O'Rahilly S, Stadler DD, Rosenfeld RG, Pratt KL, LaFranchi SH, Purnell JQ (2003a) Serum ghrelin levels are inversely correlated with body mass index, age, and insulin concentrations in normal children and are markedly increased in Prader-Willi syndrome. J Clin Endocrinol Metab 88(1):174–178

Haqq AM, Grambow SC, Muehlbauer M, Newgard CB, Svetkey LP, Carrel AL, Yanovski JA, Purnell JQ, Freemark M (2008) Ghrelin concentrations in Prader-Willi syndrome (PWS) infants and children: changes during development. Clin Endocrinol (Oxf) 69(6):911–920. doi:10.1111/j.1365-2265.2008.03385.x, CEN3385 [pii]

Haqq AM, Stadler DD, Rosenfeld RG, Pratt KL, Weigle DS, Frayo RS, LaFranchi SH, Cummings DE, Purnell JQ (2003b) Circulating ghrelin levels are suppressed by meals and octreotide therapy in children with Prader-Willi syndrome. J Clin Endocrinol Metab 88(8):3573–3576

Harrold JA, Williams G (2003) The cannabinoid system: a role in both the homeostatic and hedonic control of eating? Br J Nutr 90(4):729–734. doi:S000711450300179X, [pii]

Hillman JB, Tong J, Tschop M (2011) Ghrelin biology and its role in weight-related disorders. Discov Med 11(61):521–528

Holsen LM, Zarcone JR, Brooks WM, Butler MG, Thompson TI, Ahluwalia JS, Nollen NL, Savage CR (2006) Neural mechanisms underlying hyperphagia in Prader-Willi syndrome. Obesity (Silver Spring) 14(6):1028–1037. doi:10.1038/oby.2006.118, 14/6/1028 [pii]

Hotta M, Ohwada R, Akamizu T, Shibasaki T, Takano K, Kangawa K (2009) Ghrelin increases hunger and food intake in patients with restricting-type anorexia nervosa: a pilot study. Endocr J 56(9):1119–1128. doi:JST.JSTAGE/endocrj/K09E-168, [pii]

Hyman SE, Malenka RC, Nestler EJ (2006) Neural mechanisms of addiction: the role of reward-related learning and memory. Annu Rev Neurosci 29:565–598. doi:10.1146/annurev.neuro.29.051605.113009

Jerlhag E (2008) Systemic administration of ghrelin induces conditioned place preference and stimulates accumbal dopamine. Addict Biol 13(3–4):358–363. doi:10.1111/j.1369-1600.2008. 00125.x, ADB125 [pii]

Jerlhag E, Egecioglu E, Dickson SL, Andersson M, Svensson L, Engel JA (2006) Ghrelin stimulates locomotor activity and accumbal dopamine-overflow via central cholinergic systems in mice: implications for its involvement in brain reward. Addict Biol 11(1):45–54. doi:10.1111/j.1369-1600.2006.00002.x, ADB002 [pii]

Jerlhag E, Egecioglu E, Dickson SL, Douhan A, Svensson L, Engel JA (2007) Ghrelin administration into tegmental areas stimulates locomotor activity and increases extracellular concentration of dopamine in the nucleus accumbens. Addict Biol 12(1):6–16. doi:10.1111/j. 1369-1600.2006.00041.x, ADB041 [pii]

Jerlhag E, Egecioglu E, Dickson SL, Engel JA (2011) Glutamatergic regulation of ghrelin-induced activation of the mesolimbic dopamine system. Addict Biol 16(1):82–91. doi:10. 1111/j.1369-1600.2010.00231.x, ADB231 [pii]

Jerlhag E, Egecioglu E, Dickson SL, Svensson L, Engel JA (2008) Alpha-conotoxin MII-sensitive nicotinic acetylcholine receptors are involved in mediating the ghrelin-induced locomotor stimulation and dopamine overflow in nucleus accumbens. Eur Neuropsychopharmacol 18(7):508–518. doi:10.1016/j.euroneuro.2008.02.006. S0924-977X(08)00051-5 [pii]

Jiang H, Betancourt L, Smith RG (2006) Ghrelin amplifies dopamine signaling by cross talk involving formation of growth hormone secretagogue receptor/dopamine receptor subtype 1 heterodimers. Mol Endocrinol 20(8):1772–1785. doi:10.1210/me.2005-0084, me.2005-0084 [pii]

Kageyama H, Takenoya F, Shiba K, Shioda S (2010) Neuronal circuits involving ghrelin in the hypothalamus-mediated regulation of feeding. Neuropeptides 44(2):133–138. doi:10.1016/j. npep.2009.11.010, S0143-4179(09)00139-5 [pii]

Keen-Rhinehart E, Bartness TJ (2005) Peripheral ghrelin injections stimulate food intake, foraging, and food hoarding in Siberian hamsters. Am J Physiol Regul Integr Comp Physiol 288(3):R716–722. doi:10.1152/ajpregu.00705.2004, 00705.2004 [pii]

Kern A, Albarran-Zeckler R, Walsh HE, Smith RG (2012) Apo-ghrelin receptor forms heteromers with DRD2 in hypothalamic neurons and is essential for anorexigenic effects of DRD2 agonism. Neuron 73(2):317–332. doi:10.1016/j.neuron.2011.10.038, S0896-6273(11) 01087-7 [pii]

King SJ, Isaacs AM, O'Farrell E, Abizaid A (2011) Motivation to obtain preferred foods is enhanced by ghrelin in the ventral tegmental area. Horm Behav 60(5):572–580. doi:10.1016/j. yhbeh.2011.08.006, S0018-506X(11)00188-7 [pii]

Kirchner H, Gutierrez JA, Solenberg PJ, Pfluger PT, Czyzyk TA, Willency JA, Schurmann A, Joost HG, Jandacek RJ, Hale JE, Heiman ML, Tschop MH (2009) GOAT links dietary lipids with the endocrine control of energy balance. Nat Med 15(7):741–745. doi:10.1038/nm.1997, nm.1997 [pii]

Kola B, Farkas I, Christ-Crain M, Wittmann G, Lolli F, Amin F, Harvey-White J, Liposits Z, Kunos G, Grossman AB, Fekete C, Korbonits M (2008) The orexigenic effect of ghrelin is mediated through central activation of the endogenous cannabinoid system. PLoS ONE 3(3):e1797. doi:10.1371/journal.pone.0001797

Korotkova TM, Sergeeva OA, Eriksson KS, Haas HL, Brown RE (2003) Excitation of ventral tegmental area dopaminergic and nondopaminergic neurons by orexins/hypocretins. J Neurosci 23(1):7–11. doi:23/1/7, [pii]

Lamont EW, Patterson Z, Rodrigues T, Vallejos O, Blum ID, Abizaid A (2012) Ghrelin-deficient mice have fewer orexin cells and reduced cFOS expression in the mesolimbic dopamine pathway under a restricted feeding paradigm. Neuroscience 218:12–19. doi:10.1016/j. neuroscience.2012.05.046, S0306-4522(12)00534-9 [pii]

Landgren S, Simms JA, Thelle DS, Strandhagen E, Bartlett SE, Engel JA, Jerlhag E (2011) The ghrelin signalling system is involved in the consumption of sweets. PLoS ONE 6(3):e18170. doi:10.1371/journal.pone.0018170 le Roux CW, Patterson M, Vincent RP, Hunt C, Ghatei MA, Bloom SR (2005) Postprandial plasma ghrelin is suppressed proportional to meal calorie content in normal-weight but not obese subjects. J Clin Endocrinol Metab 90(2):1068–1071. doi:10.1210/jc.2004-1216, jc. 2004-1216 [pii]

Levin F, Edholm T, Schmidt PT, Gryback P, Jacobsson H, Degerblad M, Hoybye C, Holst JJ, Rehfeld JF, Hellstrom PM, Naslund E (2006) Ghrelin stimulates gastric emptying and hunger in normal-weight humans. J Clin Endocrinol Metab 91(9):3296–3302. doi:10.1210/jc. 2005-2638, jc.2005-2638 [pii]

Lin L, Saha PK, Ma X, Henshaw IO, Shao L, Chang BH, Buras ED, Tong Q, Chan L, McGuinness OP, Sun Y (2011) Ablation of ghrelin receptor reduces adiposity and improves insulin sensitivity during aging by regulating fat metabolism in white and brown adipose tissues. Aging Cell 10(6):996–1010. doi:10.1111/j.1474-9726.2011.00740.x

Liu B, Garcia EA, Korbonits M (2011) Genetic studies on the ghrelin, growth hormone secretagogue receptor (GHSR) and ghrelin O-acyl transferase (GOAT) genes. Peptides 32(11):2191–2207. doi:10.1016/j.peptides.2011.09.006, S0196-9781(11)00372-X [pii]

Lutter M, Sakata I, Osborne-Lawrence S, Rovinsky SA, Anderson JG, Jung S, Birnbaum S, Yanagisawa M, Elmquist JK, Nestler EJ, Zigman JM (2008) The orexigenic hormone ghrelin defends against depressive symptoms of chronic stress. Nat Neurosci 11(7):752–753. doi:10. 1038/nn.2139, nn.2139 [pii]

Ma X, Lin L, Qin G, Lu X, Fiorotto M, Dixit VD, Sun Y (2011) Ablations of ghrelin and ghrelin receptor exhibit differential metabolic phenotypes and thermogenic capacity during aging. PLoS ONE 6(1):e16391. doi:10.1371/journal.pone.0016391

Mahler SV, Smith RJ, Moorman DE, Sartor GC, Aston-Jones G (2012) Multiple roles for orexin/ hypocretin in addiction. Prog Brain Res 198:79–121. doi:10.1016/B978-0-444-59489-1. 00007-0, B978-0-444-59489-1.00007-0 [pii]

Malik S, McGlone F, Bedrossian D, Dagher A (2008) Ghrelin modulates brain activity in areas that control appetitive behavior. Cell Metab 7(5):400–409. doi:10.1016/j.cmet.2008.03.007, S1550-4131(08)00078-8 [pii]

McCallum SE, Taraschenko OD, Hathaway ER, Vincent MY, Glick SD (2011) Effects of 18-methoxycoronaridine on ghrelin-induced increases in sucrose intake and accumbal dopamine overflow in female rats. Psychopharmacology 215(2):247–256. doi:10.1007/ s00213-010-2132-0

Merkestein M, Brans MA, Luijendijk MC, de Jong JW, Egecioglu E, Dickson SL, Adan RA (2012) Ghrelin mediates anticipation to a palatable meal in rats. Obesity (Silver Spring) 20(5):963–971. doi:10.1038/oby.2011.389, oby2011389 [pii]

Miljic D, Pekic S, Djurovic M, Doknic M, Milic N, Casanueva FF, Ghatei M, Popovic V (2006) Ghrelin has partial or no effect on appetite, growth hormone, prolactin, and cortisol release in patients with anorexia nervosa. J Clin Endocrinol Metab 91(4):1491–1495. doi:10.1210/jc. 2005-2304, jc.2005-2304 [pii]

Miller JL, James GA, Goldstone AP, Couch JA, He G, Driscoll DJ, Liu Y (2007) Enhanced activation of reward mediating prefrontal regions in response to food stimuli in Prader-Willi syndrome. J Neurol Neurosurg Psychiatry 78(6):615–619. doi:10.1136/jnnp.2006.099044 jnnp.2006.099044 [pii]

Mokrosinski J, Holst B (2010) Modulation of the constitutive activity of the ghrelin receptor by use of pharmacological tools and mutagenesis. Methods Enzymol 484:53–73. doi:10.1016/ B978-0-12-381298-8.00003-4, B978-0-12-381298-8.00003-4 [pii]

Monteleone P, Fabrazzo M, Tortorella A, Martiadis V, Serritella C, Maj M (2005) Circulating ghrelin is decreased in non-obese and obese women with binge eating disorder as well as in obese non-binge eating women, but not in patients with bulimia nervosa. Psychoneuroendo-crinology 30(3):243–250. doi:10.1016/j.psyneuen.2004.07.004, S0306-4530(04)00125-8 [pii]

Monteleone P, Tortorella A, Castaldo E, Di Filippo C, Maj M (2007) The Leu72Met polymorphism of the ghrelin gene is significantly associated with binge eating disorder. Psychiatr Genet 17(1):13–16. doi:10.1097/YPG.0b013e328010e2c3, 00041444-200702000-00007 [pii]

Morpurgo PS, Resnik M, Agosti F, Cappiello V, Sartorio A, Spada A (2003) Ghrelin secretion in severely obese subjects before and after a 3-week integrated body mass reduction program. J Endocrinol Invest 26(8):723–727. doi:5567, [pii]

Nagaya N, Itoh T, Murakami S, Oya H, Uematsu M, Miyatake K, Kangawa K (2005) Treatment of cachexia with ghrelin in patients with COPD. Chest 128(3):1187–1193. doi:10.1378/chest. 128.3.1187, 128/3/1187 [pii]

Nagaya N, Moriya J, Yasumura Y, Uematsu M, Ono F, Shimizu W, Ueno K, Kitakaze M, Miyatake K, Kangawa K (2004) Effects of ghrelin administration on left ventricular function, exercise capacity, and muscle wasting in patients with chronic heart failure. Circulation 110(24):3674–3679. doi:10.1161/01.CIR.0000149746.62908.BB, 01.CIR.0000149746.62908. BB [pii]

Nakamura T, Uramura K, Nambu T, Yada T, Goto K, Yanagisawa M, Sakurai T (2000) Orexin-induced hyperlocomotion and stereotypy are mediated by the dopaminergic system. Brain Res 873(1):181–187. doi:S0006-8993(00)02555-5, [pii]

Nakazato M, Murakami N, Date Y, Kojima M, Matsuo H, Kangawa K, Matsukura S (2001) A role for ghrelin in the central regulation of feeding. Nature 409(6817):194–198

Naleid AM, Grace MK, Cummings DE, Levine AS (2005) Ghrelin induces feeding in the mesolimbic reward pathway between the ventral tegmental area and the nucleus accumbens. Peptides 26(11):2274–2279. doi:10.1016/j.peptides.2005.04.025, S0196-9781(05)00350-5 [pii]

Neary MT, Batterham RL (2010) Gaining new insights into food reward with functional neuroimaging. Forum Nutr 63:152–163. doi:10.1159/000264403, 000264403 [pii]

Neary NM, Small CJ, Wren AM, Lee JL, Druce MR, Palmieri C, Frost GS, Ghatei MA, Coombes RC, Bloom SR (2004) Ghrelin increases energy intake in cancer patients with impaired appetite: acute, randomized, placebo-controlled trial. J Clin Endocrinol Metab 89(6):2832–2836. doi:10.1210/jc.2003-031768, 89/6/2832[pii]

Ogiso K, Asakawa A, Amitani H, Inui A (2011) Ghrelin and anorexia nervosa: a psychosomatic perspective. Nutrition 27(10):988–993. doi:10.1016/j.nut.2011.05.005, S0899-9007(11) 00155-9 [pii]

Olszewski PK, Li D, Grace MK, Billington CJ, Kotz CM, Levine AS (2003) Neural basis of orexigenic effects of ghrelin acting within lateral hypothalamus. Peptides 24(4):597–602. doi:S0196978103001050, [pii]

Overduin J, Figlewicz DP, Bennett-Jay J, Kittleson S, Cummings DE (2012) Ghrelin increases the motivation to eat, but does not alter food palatability. Am J Physiol Regul Integr Comp Physiol 303(3):R259–269. doi:10.1152/ajpregu.00488.2011, ajpregu.00488.2011 [pii]

Palmiter RD (2007) Is dopamine a physiologically relevant mediator of feeding behavior? Trends Neurosci 30(8):375–381. doi:10.1016/j.tins.2007.06.004, S0166-2236(07)00133-6 [pii]

Patterson ZR, Ducharme R, Anisman H, Abizaid A (2010) Altered metabolic and neurochemical responses to chronic unpredictable stressors in ghrelin receptor-deficient mice. Eur J Neurosci 32(4):632–639. doi:10.1111/j.1460-9568.2010.07310.x, EJN7310 [pii]

Patterson ZR, Khazall R, Mackay H, Anisman H, Abizaid A (2013) Central ghrelin signaling mediates the metabolic response of C57BL/6 male mice to chronic social defeat stress. Endocrinology 154(3):1080–1091. doi:10.1210/en.2012-1834, en.2012-1834 [pii]

Perello M, Sakata I, Birnbaum S, Chuang JC, Osborne-Lawrence S, Rovinsky SA, Woloszyn J, Yanagisawa M, Lutter M, Zigman JM (2010) Ghrelin increases the rewarding value of high-fat diet in an orexin-dependent manner. Biol Psychiatry 67(9):880–886. doi:10.1016/j. biopsych.2009.10.030, S0006-3223(09)01318-3 [pii]

Perello M, Scott MM, Sakata I, Lee CE, Chuang JC, Osborne-Lawrence S, Rovinsky SA, Elmquist JK, Zigman JM (2012) Functional implications of limited leptin receptor and ghrelin receptor coexpression in the brain. J Comp Neurol 520(2):281–294. doi:10.1002/cne.22690

Perello M, Zigman JM (2012) The role of ghrelin in reward-based eating. Biol Psychiatry 72(5):347–353. doi:10.1016/j.biopsych.2012.02.016, S0006-3223(12)00143-6 [pii]

Perreault M, Istrate N, Wang L, Nichols AJ, Tozzo E, Stricker-Krongrad A (2004) Resistance to the orexigenic effect of ghrelin in dietary-induced obesity in mice: reversal upon weight loss. Int J Obes Relat Metab Disord 28(7):879–885. doi:10.1038/sj.ijo.0802640, 0802640 [pii]

Petersen PS, Woldbye DP, Madsen AN, Egerod KL, Jin C, Lang M, Rasmussen M, Beck-Sickinger AG, Holst B (2009) In vivo characterization of high Basal signaling from the ghrelin receptor. Endocrinology 150(11):4920–4930. doi:10.1210/en.2008-1638, en. 2008-1638 [pii]

Pfluger PT, Kirchner H, Gunnel S, Schrott B, Perez-Tilve D, Fu S, Benoit SC, Horvath T, Joost HG, Wortley KE, Sleeman MW, Tschop MH (2008) Simultaneous deletion of ghrelin and its receptor increases motor activity and energy expenditure. Am J Physiol Gastrointest Liver Physiol 294(3):G610–618. doi:10.1152/ajpgi.00321.2007, 00321.2007 [pii]

Purtell L, Sze L, Loughnan G, Smith E, Herzog H, Sainsbury A, Steinbeck K, Campbell LV, Viardot A (2011) In adults with Prader-Willi syndrome, elevated ghrelin levels are more consistent with hyperphagia than high PYY and GLP-1 levels. Neuropeptides 45(4):301–307. doi:10.1016/j.npep.2011.06.001, S0143-4179(11)00046-1 [pii]

Rediger A, Piechowski CL, Yi CX, Tarnow P, Strotmann R, Gruters A, Krude H, Schoneberg T, Tschop MH, Kleinau G, Biebermann H (2011) Mutually opposite signal modulation by hypothalamic heterodimerization of ghrelin and melanocortin-3 receptors. J Biol Chem 286(45):39623–39631. doi:10.1074/jbc.M111.287607, M111.287607 [pii]

Reed JA, Benoit SC, Pfluger PT, Tschop MH, D'Alessio DA, Seeley RJ (2008) Mice with chronically increased circulating ghrelin develop age-related glucose intolerance. Am J Physiol Endocrinol Metab 294(4):E752–760. doi:10.1152/ajpendo.00463.2007, 00463.2007 [pii]

Rodriguez EM, Blazquez JL, Guerra M (2010) The design of barriers in the hypothalamus allows the median eminence and the arcuate nucleus to enjoy private milieus: the former opens to the portal blood and the latter to the cerebrospinal fluid. Peptides 31(4):757–776. doi:10.1016/j. peptides.2010.01.003, S0196-9781(10)00023-9 [pii]

Sakata I, Nakano Y, Osborne-Lawrence S, Rovinsky SA, Lee CE, Perello M, Anderson JG, Coppari R, Xiao G, Lowell BB, Elmquist JK, Zigman JM (2009) Characterization of a novel ghrelin cell reporter mouse. Regul Pept 155(1–3):91–98. doi:10.1016/j.regpep.2009.04.001, S0167-0115(09)00077-9 [pii]

Sakata I, Yamazaki M, Inoue K, Hayashi Y, Kangawa K, Sakai T (2003) Growth hormone secretagogue receptor expression in the cells of the stomach-projected afferent nerve in the rat nodose ganglion. Neurosci Lett 342(3):183–186. doi:S0304394003002945, [pii]

Saper CB, Chou TC, Elmquist JK (2002) The need to feed: homeostatic and hedonic control of eating. Neuron 36(2):199–211. doi:S0896627302009698 [pii]

Sato T, Kurokawa M, Nakashima Y, Ida T, Takahashi T, Fukue Y, Ikawa M, Okabe M, Kangawa K, Kojima M (2008) Ghrelin deficiency does not influence feeding performance. Regul Pept 145(1–3):7–11. doi:10.1016/j.regpep.2007.09.010, S0167-0115(07)00186-3 [pii]

Schaeffer M, Langlet F, Lafont C, Molino F, Hodson DJ, Roux T, Lamarque L, Verdie P, Bourrier E, Dehouck B, Baneres JL, Martinez J, Mery PF, Marie J, Trinquet E, Fehrentz JA, Prevot V, Mollard P (2013) Rapid sensing of circulating ghrelin by hypothalamic appetite-modifying neurons. Proc Natl Acad Sci U S A 110(4):1512–1517. doi:10.1073/pnas. 1212137110, 1212137110 [pii]

Schellekens H, Finger BC, Dinan TG, Cryan JF (2012) Ghrelin signalling and obesity: at the interface of stress, mood and food reward. Pharmacol Ther 135(3):316–326. doi:10.1016/j. pharmthera.2012.06.004, S0163-7258(12)00122-2 [pii]

Schellekens H, van Oeffelen WE, Dinan TG, Cryan JF (2013) Promiscuous dimerization of the growth hormone secretagogue receptor (GHS-R1a) attenuates ghrelin-mediated signaling. J Biol Chem 288(1):181–191. doi:10.1074/jbc.M112.382473, M112.382473 [pii]

Schmid DA, Held K, Ising M, Uhr M, Weikel JC, Steiger A (2005) Ghrelin stimulates appetite, imagination of food, GH, ACTH, and cortisol, but does not affect leptin in normal controls. Neuropsychopharmacology 30(6):1187–1192. doi:10.1038/sj.npp.1300670, 1300670 [pii]

Schwartz MW, Woods SC, Porte D Jr, Seeley RJ, Baskin DG (2000) Central nervous system control of food intake. Nature 404(6778):661–671. doi:10.1038/35007534

Scott MM, Perello M, Chuang JC, Sakata I, Gautron L, Lee CE, Lauzon D, Elmquist JK, Zigman JM (2012) Hindbrain ghrelin receptor signaling is sufficient to maintain fasting glucose. PLoS One 7(8):e44089. doi:10.1371/journal.pone.0044089, PONE-D-12-16911 [pii]

Shimbara T, Mondal MS, Kawagoe T, Toshinai K, Koda S, Yamaguchi H, Date Y, Nakazato M (2004) Central administration of ghrelin preferentially enhances fat ingestion. Neurosci Lett 369(1):75–79. doi:10.1016/j.neulet.2004.07.060, S0304-3940(04)00943-7 [pii]

Skibicka KP, Dickson SL (2011) Ghrelin and food reward: the story of potential underlying substrates. Peptides 32(11):2265–2273. doi:10.1016/j.peptides.2011.05.016, S0196-9781(11)00208-7 [pii]

Skibicka KP, Hansson C, Alvarez-Crespo M, Friberg PA, Dickson SL (2011) Ghrelin directly targets the ventral tegmental area to increase food motivation. Neuroscience 180:129–137. doi:10.1016/j.neuroscience.2011.02.016, S0306-4522(11)00157-6 [pii]

Strasser F, Lutz TA, Maeder MT, Thuerlimann B, Bueche D, Tschop M, Kaufmann K, Holst B, Brandle M, von Moos R, Demmer R, Cerny T (2008) Safety, tolerability and pharmacokinetics of intravenous ghrelin for cancer-related anorexia/cachexia: a randomised, placebo-controlled, double-blind, double-crossover study. Br J Cancer 98(2):300–308. doi:10.1038/sj.bjc.6604148, 6604148 [pii]

Stratford TR, Kelley AE (1999) Evidence of a functional relationship between the nucleus accumbens shell and lateral hypothalamus subserving the control of feeding behavior. J Neurosci 19(24):11040–11048

Sun Y, Ahmed S, Smith RG (2003) Deletion of ghrelin impairs neither growth nor appetite. Mol Cell Biol 23(22):7973–7981

Sun Y, Butte NF, Garcia JM, Smith RG (2008) Characterization of adult ghrelin and ghrelin receptor knockout mice under positive and negative energy balance. Endocrinology 149(2):843–850. doi:10.1210/en.2007-0271, en.2007-0271 [pii]

Sun Y, Wang P, Zheng H, Smith RG (2004) Ghrelin stimulation of growth hormone release and appetite is mediated through the growth hormone secretagogue receptor. Proc Natl Acad Sci U S A 101(13):4679–4684. doi:10.1073/pnas.0305930101, 0305930101 [pii]

Suzuki K, Simpson KA, Minnion JS, Shillito JC, Bloom SR (2010) The role of gut hormones and the hypothalamus in appetite regulation. Endocr J 57(5):359–372. doi:JST.JSTAGE/endocrj/K10E-077, [pii]

Tan TM, Vanderpump M, Khoo B, Patterson M, Ghatei MA, Goldstone AP (2004) Somatostatin infusion lowers plasma ghrelin without reducing appetite in adults with Prader-Willi syndrome. J Clin Endocrinol Metab 89(8):4162–4165. doi:10.1210/jc.2004-0835, 89/8/4162 [pii]

Tanaka M, Nakahara T, Kojima S, Nakano T, Muranaga T, Nagai N, Ueno H, Nakazato M, Nozoe S, Naruo T (2004) Effect of nutritional rehabilitation on circulating ghrelin and growth hormone levels in patients with anorexia nervosa. Regul Pept 122(3):163–168. doi:10.1016/j.regpep.2004.06.015, S0167011504002058 [pii]

Tanaka M, Naruo T, Yasuhara D, Tatebe Y, Nagai N, Shiiya T, Nakazato M, Matsukura S, Nozoe S (2003) Fasting plasma ghrelin levels in subtypes of anorexia nervosa. Psychoneuroendocrinology 28(7):829–835. doi:S0306453002000665, [pii]

Toshinai K, Date Y, Murakami N, Shimada M, Mondal MS, Shimbara T, Guan JL, Wang QP, Funahashi H, Sakurai T, Shioda S, Matsukura S, Kangawa K, Nakazato M (2003) Ghrelin-induced food intake is mediated via the orexin pathway. Endocrinology 144(4):1506–1512

Tschop M, Weyer C, Tataranni PA, Devanarayan V, Ravussin E, Heiman ML (2001) Circulating ghrelin levels are decreased in human obesity. Diabetes 50(4):707–709

Verhagen LA, Egecioglu E, Luijendijk MC, Hillebrand JJ, Adan RA, Dickson SL (2011) Acute and chronic suppression of the central ghrelin signaling system reveals a role in food anticipatory activity. Eur Neuropsychopharmacol 21(5):384–392. doi:10.1016/j.euroneuro.2010.06.005, S0924-977X(10)00118-5 [pii]

Walker AK, Ibia IE, Zigman JM (2012) Disruption of cue-potentiated feeding in mice with blocked ghrelin signaling. Physiol Behav 108:34–43. doi:10.1016/j.physbeh.2012.10.003, S0031-9384(12)00327-7 [pii]

Weinberg ZY, Nicholson ML, Currie PJ (2011) 6-Hydroxydopamine lesions of the ventral tegmental area suppress ghrelin's ability to elicit food-reinforced behavior. Neurosci Lett 499(2):70–73. doi:10.1016/j.neulet.2011.05.034, S0304-3940(11)00648-3 [pii]

Willesen MG, Kristensen P, Romer J (1999) Co-localization of growth hormone secretagogue receptor and NPY mRNA in the arcuate nucleus of the rat. Neuroendocrinology 70(5):306–316. doi:nen70306, [pii]

Williams KW, Elmquist JK (2012) From neuroanatomy to behavior: central integration of peripheral signals regulating feeding behavior. Nat Neurosci 15(10):1350–1355. doi:10.1038/nn.3217, nn.3217 [pii]

Wortley KE, del Rincon JP, Murray JD, Garcia K, Iida K, Thorner MO, Sleeman MW (2005) Absence of ghrelin protects against early-onset obesity. J Clin Invest 115(12):3573–3578. doi:10.1172/JCI26003

Wren AM, Seal LJ, Cohen MA, Brynes AE, Frost GS, Murphy KG, Dhillo WS, Ghatei MA, Bloom SR (2001) Ghrelin enhances appetite and increases food intake in humans. J Clin Endocrinol Metab 86(12):5992

Wu Q, Palmiter RD (2011) GABAergic signaling by AgRP neurons prevents anorexia via a melanocortin-independent mechanism. Eur J Pharmacol 660(1):21–27. doi:10.1016/j.ejphar.2010.10.110, S0014-2999(10)01275-6 [pii]

Wynne K, Giannitsopoulou K, Small CJ, Patterson M, Frost G, Ghatei MA, Brown EA, Bloom SR, Choi P (2005) Subcutaneous ghrelin enhances acute food intake in malnourished patients who receive maintenance peritoneal dialysis: a randomized, placebo-controlled trial. J Am Soc Nephrol 16(7):2111–2118. doi:10.1681/ASN.2005010039, ASN.2005010039 [pii]

Yang N, Liu X, Ding EL, Xu M, Wu S, Liu L, Sun X, Hu FB (2009) Impaired ghrelin response after high-fat meals is associated with decreased satiety in obese and lean Chinese young adults. J Nutr 139(7):1286–1291. doi:10.3945/jn.109.104406, jn.109.104406 [pii]

Zhao TJ, Liang G, Li RL, Xie X, Sleeman MW, Murphy AJ, Valenzuela DM, Yancopoulos GD, Goldstein JL, Brown MS (2010) Ghrelin O-acyltransferase (GOAT) is essential for growth hormone-mediated survival of calorie-restricted mice. Proc Natl Acad Sci U S A 107(16):7467–7472. doi:10.1073/pnas.1002271107, 1002271107 [pii]

Zheng H, Patterson LM, Berthoud HR (2007) Orexin signaling in the ventral tegmental area is required for high-fat appetite induced by opioid stimulation of the nucleus accumbens. J Neurosci 27(41):11075–11082. doi:10.1523/JNEUROSCI.3542-07.2007, 27/41/11075 [pii]

Zigman JM, Jones JE, Lee CE, Saper CB, Elmquist JK (2006) Expression of ghrelin receptor mRNA in the rat and the mouse brain. J Comp Neurol 494(3):528–548. doi:10.1002/cne.20823

Zigman JM, Nakano Y, Coppari R, Balthasar N, Marcus JN, Lee CE, Jones JE, Deysher AE, Waxman AR, White RD, Williams TD, Lachey JL, Seeley RJ, Lowell BB, Elmquist JK (2005) Mice lacking ghrelin receptors resist the development of diet-induced obesity. J Clin Invest 115(12):3564–3572. doi:10.1172/JCI26002

The Ghrelin Receptor: A Novel Therapeutic Target for Obesity

Harriët Schellekens, Timothy G. Dinan and John F. Cryan

Abstract The obesity epidemic has evolved into an ever expanding serious global health concern. Several physiological as well as environmental factors have contributed to the rise in obesity incidence. Obesity or being overweight results from an energy imbalance characterized by an excess of caloric intake more often than not combined with a reduced energy expenditure, for example, due to physical inactivity. Nutrient status is communicated via circulating gut hormones, which all act on the brain to regulate short-and long-term appetite and the body's metabolism and this brain-gut axis communication is dysregulated under metabolic conditions, such as obesity. Ghrelin is the only peripheral-derived hormone, which exerts an orexigenic effect via the modulation of central circuitries, and has therefore received considerable focus in the pharmaceutical industry for the development of anti-obesity therapeutics. Two subtypes of the ghrelin receptor have been reported to date, the growth hormone secretagogue (GHS-R1a) receptor 1a isoform, which is activated by acylated ghrelin, and the truncated isoform GHS-R1b, which is

H. Schellekens · T. G. Dinan · J. F. Cryan
Food for Health Ireland, University College Cork, Cork, Ireland
e-mail: H.Schellekens@ucc.ie

T. G. Dinan
e-mail: T.Dinan@ucc.ie

T. G. Dinan · J. F. Cryan
Laboratory of Neurogastroenterology, Alimentary Pharmabiotic Centre,
University College Cork, Cork, Ireland

T. G. Dinan
Department of Psychiatry, University College Cork, Cork, Ireland

J. F. Cryan (✉)
Deparment of Anatomy and Neuroscience, Western Gateway Building,
University College Cork, Cork, Ireland
e-mail: J.Cryan@ucc.ie

H. Schellekens
School of Pharmacy, University College Cork, Cork, Ireland

J. Portelli and I. Smolders (eds.), *Central Functions of the Ghrelin Receptor*,
The Receptors 25, DOI: 10.1007/978-1-4939-0823-3_6,
© Springer Science+Business Media New York 2014

functionally inactive. Interestingly, the GHS-R1b receptor has been shown to exert a dominant-negative effect on GHS-R1a receptor functioning via the formation of a GHS-R1a/1b dimer. The GHS-R1a is expressed in multiple brain regions with ghrelin's orexigenic effect on homeostatic food intake being mainly mediated in the arcuate nucleus of the hypothalamus and hedonic aspects being mediated via GHS-R1a receptors in the mesolimbic dopaminergic circuitry. In this review, we discuss the role of ghrelin in the hypothalamic regulation of appetite and highlight the additional dimension of the ghrelin/GHS-R1a receptor axis as a target in obesity via manipulation of the ghrelin-mediated nonhomeostatic rewarding aspect of food intake behavior. In addition, we review the current understanding of the role of the ghrelin receptor isoforms as targets in obesity and discuss the potential of hetero-dimers in the development of more specific anti-obesity therapeutics.

Keywords Obesity · Ghrelin · Growth hormone secretagogue receptor · Food intake behavior · Heterodimerization

Introduction

The incidence of obesity continues to increase globally and has reached epidemic proportions. The rise in obesity prevalence also concomitantly increases obesity-associated comorbidities and represents a serious public health concern and a heavy burden on health care costs (Bloom et al. 2008). Worldwide obesity has doubled since 1980, and although there are some signs of slowing down or reaching plateau levels, an unmet need exists for novel anti-obesity pharmaco-therapies (Flegal et al. 2012; Ogden et al. 2012). According to reports from the World Health Organization (WHO), at least 2.8 million people are dying each year as a result of being overweight or obese (World_Health_Organisation 2013). In addition, the WHO estimates that within Europe, obesity is responsible for up to 8 % of total health costs and over 10 % of deaths and thus continues to be a major public health threat. Moreover, the U.S Center for Disease Control and Prevention has predicted that by 2030, about 32 million more Americans will become obese, which will constitute an obesity rate of approximately 42 % within the U.S. population (Center_for_Disease_Control 2013).

Being overweight is defined as having a body mass index (BMI), comparing body weight and height, of between 25 and 30 kg/m^2 while obesity is associated with a BMI of greater than 30 kg/m^2 (Eknoyan 2008). Obesity is often associated with comorbid diseases and complications, which have a significant adverse effect on health and decrease life expectancy (Cheng and Leiter 2006; Mikhail 2009). These obesity-associated comorbidities include insulin resistance or diabetes mellitus type II, glucose intolerance, dyslipidemia, atherosclerosis, hypertension, and a general pro-inflammatory phenotype, which all together constitute a related group of coinciding factors under the umbrella term of the Metabolic Syndrome.

Obesity is a multifactorial disease, involving behavioral, metabolic, environmental, and genetic factors as well as cultural influences and factors from socioeconomic status, making it a complex health issue to address. Being overweight or obese result from an excess accumulation of body fat or adiposity, brought about following excess consumption of high caloric foods (i.e., hyperphagia), which is also often paralleled by a decrease in energy expenditure due to insufficient physical activity, leading to an imbalance in overall energy homeostasis (Chakrabarti 2009). It is hypothesized that animals, including humans, are primed by genomic evolutionary pressure to respond to environmental cues in favor of energy intake instead of expenditure, ensuring that appropriate levels of energy are maintained during unpredictable food scarcity. Therefore, a particular contribution to the obese epidemic is been given to the substantially changed food environment from one of scarcity to one of abundance, which we have experienced over the past half century. The deposition of energy in the form of fat in adipose tissue has evolutionarily been beneficial in the wild to sustain survival, but is no longer required in modern day society as the availability of palatable foods, which are, high in calories in the form of fats and sugar, has increased dramatically in the developed nations.

While the link between the metabolic syndrome and its associated comorbidities, in particular obesity, and increased intake of unhealthy foods is obvious, current anti-obesity strategies have a low success rate, mainly due to poor adherence and commitment to diet (Bloom et al. 2008; Halford et al. 2010; Derosa and Maffioli 2012; Kang and Park 2012; Nguyen et al. 2012). Likewise, most anti-obesity pharmacotherapeutics have a low potency and efficacy and are associated with side effects and have been withdrawn from the market. Thus, novel strategies are required to curb food cravings and reduce appetite.

A potential explanation for the lack of effective anti-obesity therapeutics can be likely found in the vast redundant appetite and satiety signaling pathways regulating body weight homeostasis. Food intake is mediated by intricate peripheral and central signaling mechanisms many of which share overlapping functionalities and can compensate for one another. Therefore, the effectiveness of each pharmacotherapy targeting one particular system depends on metabolic status and both short- and long-term appetite signals. Within the intricate network of satiety signaling pathways, the acylated 28-amino acid peptide ghrelin is the first and only known peripherally produced hormone that exerts an orexigenic effect on food intake via centrally activated mechanisms (Kojima et al. 1999, 2004; Tschop et al. 2000; Nakazato et al. 2001). Therefore, targeting the central ghrelinergic system represents an attractive target for the development of novel anti-obesity pharmacotherapies (Zorrilla et al. 2006; Leite-Moreira and Soares 2007; Moulin et al. 2007; Soares et al. 2008; Chollet et al. 2009; Lu et al. 2009). The ghrelinergic system plays a prominent role in the central nervous system, and in addition, to its function in the stimulation of GH secretion, ghrelin and its receptor, the growth hormone secretagogue (GHS-R1) receptor have been shown to regulate multiple aspects of food intake behavior (Howard et al. 1996; Guan et al. 1997; Zigman et al. 2006).

This chapter will review the current understanding of the role of both ghrelin receptor isoforms, the GHS-R1a isoform, which is activated by acylated ghrelin, and the truncated isoform GHS-R1b, which is functionally inactive and their role in the hypothalamic regulation of energy homeostasis. In addition, we will discuss the role of the ghrelin/GHS-R1a receptor axis in the nonhomeostatic rewarding aspect of food intake behavior. Finally, we will review the current understanding of the ghrelinergic system as a target in obesity and the potential of heterodimers in the development of more specific anti-obesity therapeutics.

Central Circuitries of Ghrelin Signaling in Food Intake Behavior

Hypothalamic Regulation of Appetite

The regulation of appetite is mediated by a complex network of central and peripheral hormones, peptides, and receptors (Stanley et al. 2005; Schellekens et al. 2013a) mediating a bidirectional communication between the gastrointestinal tract and the central nervous system. The interaction between these peripheral and central signals regulates the homeostatic energy balance of energy intake versus energy expenditure. The intake of food and ingested nutrients triggers the peripheral release of a plethora of gut hormones and peptides, which coordinate appetite and satiety through their effects on the brain (Fig. 1a). The continued crosstalk between peripheral organs and the brain maintains a constant energy supply for cellular functions and protects against periods of food scarcity. An impairment within this two-way brain-gut axis communication (Konturek et al. 2004) may result in disorders of feeding behavior and weight gain (obesity) (Bloom et al. 2008; Swinburn et al. 2011) or weight loss (anorexia and cachexia) (Tisdale 1997; Evans et al. 2008; Dostalova and Haluzik 2009; Ogiso et al. 2011; Kaye et al. 2013). Circulating peripheral hormones, including ghrelin, reach the brain via the general circulation or via the vagus nerve (Venkova and Greenwood-Van Meerveld 2008), and relay information on satiety, adiposity, and caloric intake to the brain via the hypothalamus and the brainstem, respectively (Ahima and Antwi 2008; Simpson et al. 2008, 2009; Blevins and Baskin 2010) (Fig. 1b).

The hypothalamus is strategically positioned because it is not fully isolated by the blood–brain barrier, and can interact with signals from the periphery via the blood circulation. An additional pathway for integration of peripheral signals to the hypothalamus is provided via projections from the brainstem, which receives vagal afferents from the periphery to the area postrema in the brainstem (Venkova and Greenwood-Van Meerveld 2008). This represents the major neuroanatomical connection providing communication between the gastrointestinal (GI) tract and the brain. The hypothalamus can therefore be considered as the main processor and integrator of peripheral metabolic information controlling food intake and plays a

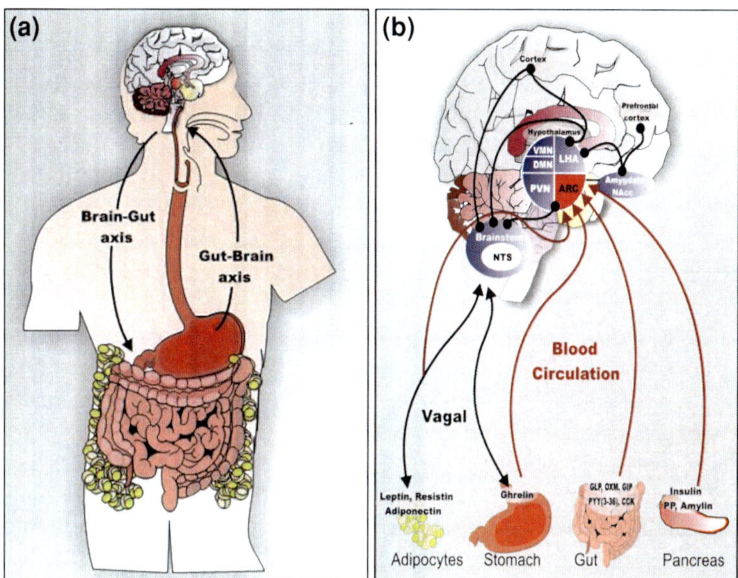

Fig. 1 *The bidirectional brain-gut axis in energy homeostasis.* Hormones produced in the gut signal information on nutrient status to the brain (*gut-brain axis*), which integrates these gut hormones and neurotransmitters to regulate food intake and energy metabolism via feedback to the periphery (*brain-gut axis*) (**a**). Peripheral signals enter the brain via the vagus nerve or via the blood circulation after crossing the BBB. Projections exists from the *ARC* to other hypothalamic nuclei and to other brain areas involved in the regulation of food intake (**b**). Abbreviations: *ARC* arcuate nucleus; *BBB* blood barin barrier; *CCK* cholecystokinin; *DMN* dorsal medial nucleus; *GIP* glucose-dependent insulinotropic polypeptide (gastric inhibitory peptide); *GLP* glucagon-like peptide 1; *LHA* lateral hypothalamic area; *NAcc* nucleus accumbens; *NTS* nucleus of the solitary tract (nucleus tractus solitarii); *OXM* oxyntomodulin; *PP* pancreatic polypeptide; *PVN* paraventricular nucleus; *PYY* peptide tyrosine-tyrosine; *VMN* ventromedial nucleus

key role in the central regulation of appetite and energy metabolism (Simpson et al. 2009; Suzuki et al. 2010). The major hypothalamic site controlling appetite is the arcuate nucleus (ARC), located above the median eminence, which represents the primary signaling site for peripheral satiety hormones (Blevins and Baskin 2010).

The integration and processing of peripheral peptides within the brain and subsequent brain-gut feedback can have three broad outcomes: meal termination, inhibitory modulation of intake in subsequent meals, or orexigenic effects (Schellekens et al. 2013a). The majority of gut peptides have an anorexigenic effect via central actions and include the short acting cholecystokinin (CCK), amylin, and pancreatic glucagon, which are all involved in meal termination. CCK, which is released postprandially, was the first gut hormone which demonstrated to affect food intake (Gibbs et al. 1973). Amylin, secreted from the pancreatic A-cell secretory vesicles in response to food, has its anorexigenic action via action on the hindbrain area postrema and central nucleus of the amygdala (Lutz 2006). Peptide tyrosine-tyrosine (PYY), glucagon-like peptide 1 (GLP-1), gastric inhibitory polypeptide (GIP), and oxyntomodulin (OXM) are peptides with longer term inhibitory actions on feeding. Leptin, adiponectin, and resistin are adipokines secreted from adipose tissue in proportion to fat mass and have centrally mediated effects via the hypothalamus on energy expenditure, food intake, and appetite (Zhang et al. 1994; Tovar et al. 2005; Ahima and Lazar 2008; Kadowaki et al. 2008). Leptin is expressed from the *ob* gene and secreted in proportion to fat mass (Zhang et al. 1994). The inhibition of food intake mediated by leptin is more rapid compared to the slower suppression of inter meal appetite mediated by PYY and has a longer term effect in contrast to the rapid and short acting inhibition mediated via CCK. Moreover, leptin counteracts the effect of neuropeptide Y (NPY) in the hypothalamus and stimulates synthesis of the appetite suppressant, α-melanocyte-stimulating hormone (α-MSH). Finally, pancreas-derived insulin, which is secreted proportional to fat mass and following an increase in glucose load, has equally profound effects on appetite and reaches the brain via receptor-mediated transport across the blood–brain barrier.

The 28-amino acid peptide ghrelin, aptly coined the "hunger hormone," is secreted from the stomach and is notably the only identified peripheral signal so far to act in the hypothalamus to stimulate food intake (Kojima et al. 1999; Nakazato et al. 2001). Ghrelin has a potent appetite-enhancing effect, resulting in a significant increase of food intake and has shortened meal intervals following peripheral and central administration of ghrelin in rodents (Nakazato et al. 2001; Kojima and Kangawa 2002). In addition, a potent orexigenic effect of ghrelin in humans has also been documented (Wren et al. 2001). The secretion of gastric ghrelin is enhanced in-between meals and under conditions of negative energy balance, such as fasting, starvation, and anorexia and decreases post food ingestion (Tschop et al. 2001a; Lawrence et al. 2002; Sun et al. 2004). Thus, a substantial body of evidence supports ghrelin's key role in mealtime hunger and meal initiation, increasing food intake and adiposity (Tschop et al. 2000; Cummings et al. 2001).

Central Ghrelin Signaling in Appetite Regulation

The orexigenic peptide ghrelin is predominantly synthesized in the stomach by the enteroendocrine cellular system, also referred to as ghrelin cells or Gr cells (Date et al. 2000). The ghrelin-secreting endocrine mucosal cells are named X/A like stomach cells in rat and P/D cells in humans (Date et al. 2000; Rindi et al. 2002; Sakata et al. 2002). The mature human ghrelin peptide (Fig. 2a) is encoded by the ghrelin gene (*ghrl*), which spans 7.2 kb of genomic DNA, across 6 exons located on chromosome 3p25–26 (Seim et al. 2007; Schellekens et al. 2009). The mature ghrelin hormone is enzymatically cleaved from preproghrelin and requires a posttranslational modification of an n-octanoylation mediated by the enzyme, ghrelin O-acyltransferase (GOAT), on the third serine residue to be biologically active (Gualillo et al. 2008; Gutierrez et al. 2008; Yang et al. 2008a, b).

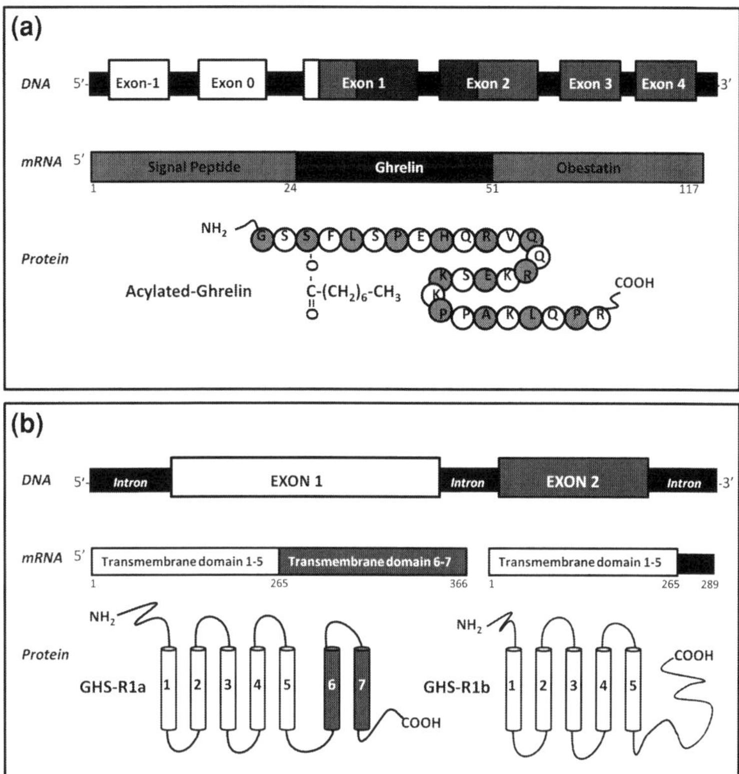

Fig. 2 *The human gastric-derived peptide ghrelin and the growth secretagogue hormone (GHS-R1) receptor*. The ghrelin gene is located on chromosome 3p25–26, composed of 6 exons and 5 introns, and transcribes the ghrelin mRNA, which is translated into the 117 amino acid protein, prepro-ghrelin. Prepro-ghrelin is processed into obestatin and the 28 amino acid mature ghrelin. The ghrelin hormone is activated and fully functional following posttranslational modification and addition of a unique octanoyl group on the third serine residue (**a**). The human full-length, GHS-R1a is processed from mRNA encoded from exon 1 and 2, located on chromosome 3p26.2. The GHS-R1b isoform results from alternative splicing of the mRNA and is encoded by exon 1 and part of the intronic sequence with an alternative polyadenylation site (adapted from Schellekens et al. 2009)

Peripheral ghrelin translates information about nutrients from the gut to the brain to determine meal initiation, meal frequency, and long-term regulation of body weight (Kojima et al. 1999; Cummings and Shannon 2003). Ghrelin plasma concentrations increase before meals and during fasting, initiating hunger and decrease after ingestion of food (Cummings et al. 2001, 2002a, b; Tschop et al. 2001b). Interestingly, plasma levels of ghrelin are significantly reduced in patients following total gastrectomy, reinforcing the stomach as the major source of circulating ghrelin (Hosoda et al. 2003). Ghrelin's effect on the central regulation of appetite and food intake are mediated via activation of the centrally expressed growth hormone secretagogue (GHS-R1a) receptor 1a, the biological functional variant of the ghrelin receptor (for review see Tamura et al. 2002; Schellekens et al. 2009, 2012; Castaneda et al. 2010; Miwa et al. 2010; Andrews 2011). The fully functional GHS-R1a is a G-protein coupled receptor (GPCR) with a 7 transmembrane domain structure spanning 366 amino acids with a molecular weight of 41 kDa (Fig. 2b). A second variant of the ghrelin receptor exists, the GHS-R1b receptor, which is a truncated polypeptide spanning only 289 amino acids and which lacks the last 2 transmembrane domains typical for GPCRs (for review see Schellekens et al. 2009) (Fig. 2b). The GHS-R1b receptor does not bind ghrelin but attenuates GHS-R1a signaling when expressed as a heterodimer (Chan and Cheng 2004; Schellekens et al. 2013b).

The central GHS-R1a receptors orchestrating ghrelin's orexigenic effect are mainly located in the hypothalamus and the brainstem (Nakazato et al. 2001; Olszewski et al. 2003b; Currie et al. 2005; Andrews 2011). The ghrelin-mediated hyperphagic effects following central administration are well documented (Melis et al. 2002; Faulconbridge et al. 2003; Olszewski et al. 2003a). Indeed, within the brain the GHS-R1a receptor has the highest expression levels in the hypothalamus, which is in line with ghrelin's role in the homeostatic regulation of energy balance (Zigman et al. 2006). Interestingly, hypothalamic GHS-R expression has been shown to be increased following fasting or chronic food restriction, reinforcing a key role for the GHS-R1a in the regulation of food intake (Kurose et al. 2005).

The hypothalamus, more specifically the ARC, coordinates the processing of peripheral cues, including ghrelin (Kohno et al. 2003). The key role of the ARC in ghrelinergic signaling was demonstrated following chemical ablation of the ARC and following antisense GHS-R1a, which completely blocked the orexigenic effect of centrally administered ghrelin (Shuto et al. 2002; Tamura et al. 2002). Within the ARC, energy balance, food intake and appetite are regulated predominantly via two neuronal populations. The first group of appetite regulating neurons is the medially located neurons containing the orexigenic NPY and agouti-related peptide (AgRP). Ghrelin activation of the GHS-R1a receptor expressed on these NPY neurons increases appetite and stimulates food intake via direct GHS-R1a mediated expression of NPY and AgRP from arcuate NPY/AgRP neurons and induce the neuronal activity markers c-Fos and Egr-1 (Chen et al. 2004; Andrews et al. 2008). This downstream signaling via NPY/AgRP neurons is confirmed by the observation that ghrelin's orexigenic effect is attenuated with specific antiserum against NPY and AgRP and following pretreatment with a nonpeptide NPY Y1 receptor antagonist. This reinforces the ghrelin-induced orexigenic effects to be mediated via downstream NPY/AgRP neurons in the ARC.

The second population of neurons are located laterally and express the anorexigenic peptides cocaine amphetamine-regulated transcript (CART) and proopiomelanocortin (POMC) and are satiety-inducing upon firing. Upon activation of the satiety pathway, POMC is processed into several peptides, including α-MSH following cleavage by prohormone convertases 1 and 2 (PC1 and PC2). The POMC-derived melanocortins bind to downstream melanocortin receptor 4 receptor (MC_4) and melanocortin 3 receptor (MC_3) to inhibit food intake (Adan et al. 2006; Marston et al. 2011; Pandit et al. 2011; Xu et al. 2011). Ghrelin also indirectly inhibits POMC/CART neurons, which do not express the GHS-R1a receptor (Cowley et al. 2003), via local projection of ghrelin-mediated GABA release from the NPY/AgRP neurons (Chen et al. 2004; Andrews et al. 2008). Thus, activation of centrally expressed GHS-R1a receptor in the ARC of the hypothalamus leads to neuronal excitation of NPY neurons and subsequent inhibition of downstream signaling via melanocortin receptors (MC_3 and MC_4) expressed in the paraventricular nucleus (PVN) and the lateral hypothalamic area (LHA) (Schellekens et al. 2009, 2012). In addition, ghrelin-mediated expression of AgRP contributes to the increase in food intake through inverse agonism of the constitutively active MC_3 and MC_4 receptors, and α-MSH antagonism, which both decrease satiety (Tolle and Low 2008). The ghrelin-mediated activation of AgRP neurons may potentially function to suppress melanocortin receptor activity during fasting in an attempt to compensate for a negative energy balance (Adan et al. 2006). Interesting to note is that ghrelin acts in an opposite manner to that of adipose-derived leptin, which increases POMC expression and POMC-derived α-MSH, while decreasing NPY and AgRP expression (Obici 2009). Finally, the GHSR1a-expressing NPY neurons project to orexin (ORX) neurons of the lateral hypothalamus to stimulate food intake. Moreover, the GHS-R1a receptor is also expressed on ORX neurons and ghrelin microinjected into the LHA can directly activate ORX neurons and increase food intake (Lawrence et al. 2002; Olszewski et al. 2003b; Yamanaka et al. 2003). Furthermore, in the hypothalamus of *ob/ob* mice expression of ORX is reduced despite upregulated NPY levels, which may potentially act as a counter regulatory system necessary to limit the adverse effects of enhanced NPY on food intake and body weight in this mouse model of obesity (Stricker-Krongrad et al. 2002). Peripherally produced ghrelin exerts its appetite-inducing effects centrally after passing through the blood–brain barrier (Banks et al. 2002, 2008; Schaeffer et al. 2013). Peripheral ghrelin also reaches the brain stem via vagal afferents to the nucleus of the solitary tract (NTS) in the, which has further projections to the ARC (but also see Arnold et al. 2006). There is also evidence for the synthesis of the neuropeptide ghrelin in the brain, albeit at a much lower levels, in specific neuronal cells of the hypothalamus (Kojima et al. 2001; Lu et al. 2002; Cowley et al. 2003; Sato et al. 2005; Schellekens et al. 2013c). Ghrelin-immunoreactivity was demonstrated in the ependymal layer of the third ventricle and between neurons of the dorsal medial nucleus (DMN), the ventromedial nucleus (VMN), the PVN, and in the ARC of the hypothalamus (Cowley et al. 2003; Hou et al. 2006). In addition, efferent projections from these ghrelin-expressing neurons to key hypothalamic circuits were shown, including to NPY neurons, POMC, CART neurons, and corticotrophin-releasing hormone (CRH) containing neurons. Moreover, ghrelin-immunoreactive neurons were found to have terminals on ORX fibers in the LHA, further linking ghrelin and ORX in the regulation of feeding behavior and energy homeostasis (Toshinai et al. 2003). Together this may suggest that ghrelinergic neurons represent a key regulatory circuit controlling energy homeostasis via the stimulation of orexigenic peptides and neurotransmitters release. The existence of ghrelin producing neurons in the

ARC was confirmed by a transgenic ghrelin-EGFP mouse model (Kageyama et al. 2008) and a more recent study demonstrated increases in ghrelin mRNA levels in a hypothalamic cell line following a dexamethasone challange (Kageyama et al. 2012). Moreover, ghrelin expression was also demonstrated in hypothalamic nuclei of humans (Montoya-Flores et al. 2012). Thus, both peripheral and central produced ghrelin constitute natural ligands for the GHS-R1a receptor. Noteworthy, the expression of ghrelin in the brain remains highly controversial as significant amounts of authentic ghrelin were absent in rodent neuronal cells and GHS-R1a receptor-expressing neurons did not receive synaptic inputs from ghrelin-immunoreactive nerve terminals in these species, suggesting considerable inconsistence between different studies (Furness et al. 2011). Nevertheless, appetite signaling via centrally expressed GHS-R1a receptors has been extensively documented and the ghrelinergic system therefore represents an excellent target for the development of anti-obesity therapeutics.

Ghrelin's Effect on Hedonic Food Intake

Recent evidence has emerged demonstrating that the orexigenic gastrointestinal hormone ghrelin is also involved in the regulation of the pleasurable and motivational aspects of food intake and in reward-based eating behavior (for review see Dickson et al. 2011; Egecioglu et al. 2011; Skibicka and Dickson 2011; Schellekens et al. 2012, 2013a). Palatable and often caloric foods (i.e., high sucrose and/or high fat) are strong reinforcers and the biggest contributor to nonhomeostatic feeding. Palatable foods stimulate rewarding pleasurable signals, which independently of metabolic needs override satiety and stimulate hedonic eating behavior and the decision to eat, leading to excess food consumption and hence, obesity. The activation of the rewarding and motivational drive to eat is equally impacted via the classical feeding peptides regulating homeostatic control of food intake, via connections to dopamine neurons in the mesolimbic circuitry (Hoebel 1985; Wise and Rompre 1989; Wise 2006; Volkow et al. 2010, 2012; Parylak et al. 2011). Indeed, metabolic hormones, including ghrelin, have been shown to be closely associated with the mesolimbic dopaminergic neurocircuitries processing the hedonic and rewarding properties of food (Narayanan et al. 2010). The GHS-R receptors are highly expressed in the suprachiasmatic, anterior hypothalamic, paraventricular, anteroventral preoptic, and tuberomammillary nuclei of the hypothalamus and in extra-hypothalamic areas of the brain, including the substantia nigra, dorsal and median raphe nucleus, hippocampus, and amygdala (Zigman et al. 2006). In addition, the GHS-R1a receptor is highly expressed on dopamine neurons of the ventral tegmental area (VTA) and nucleus accumbens (NAcc), which reinforces the involvement of ghrelin signaling in hedonic eating behavior mediated in the midbrain neurons of the mesolimbic reward system (Guan et al. 1997; Abizaid 2009; Skibicka and Dickson 2011). The mesolimbic dopaminergic projections from the VTA terminate in the ventral striatum and the prefrontal cortex, which represents the key neurocircuitry mediating anticipatory food-reward and food-seeking behavior (Richardson and Gratton 1998; Bassareo and Di Chiara 1999). Ghrelin was shown to robustly activate the VTA and NAcc following direct injection, and to robustly stimulate an eating response, supporting ghrelin's role in the dopaminergic VTA-mediated reward signaling (Naleid et al. 2005). Moreover, intra-VTA administration of ghrelin was shown to increase the consumption of rewarding foods in mice (Egecioglu et al. 2010). Interestingly, the NAcc has been disputed as a direct target for ghrelin's action on food-motivated behavior, as the motivation to work for a sucrose reward was reduced following blockade of the GHS-R1a receptor in the VTA only, while fasting-induced chow hyperphagia was unaffected (Skibicka and Dickson 2011). This data suggest that ghrelin modulates appetite and satiety via the NAcc and that the VTA affects the ghrelin-mediated rewarding properties of food, the selection of rewarding foods, and food-motivated behavior, but not ghrelin-induced chow intake (Dickson et al. 2011; Skibicka and Dickson 2011).

The ability of ghrelin to alter food reward in response to palatable foods is suggested to be mediated following dopamine release from VTA-derived dopaminergic neurons projecting to the NAcc (Jerlhag et al. 2007; Dickson et al. 2011; Skibicka and Dickson 2011). Moreover, peripheral and central ghrelin administration has been shown to directly enhance ghrelin-mediated activation of dopaminergic neurons within the VTA (Abizaid et al. 2006). This dopaminergic activation was shown to be crucial for the potentiation of food reward, as the ghrelin-mediated response on food-reinforced behavior in progressive ratio responding was absent following administration of the dopaminergic neurotoxin 6-hydroxydopamine (6-OHDA). Thus, the ghrelinergic neurocircuitries connected with the mesolimbic dopamine pathway mediate the hedonic and rewarding aspects of food and can promote the predisposition to overeat when presented with palatable and energy dense food sources.

Administration of peripheral or central ghrelin was shown to enhance hedonic feeding associated with food palatability (Shimbara et al. 2004; Disse et al. 2010). This enhanced preference for rewarding foods was shown to be dependent on the GHS-R1a receptor as it was absent in GHS-R1a knockout mice (Disse et al. 2010) or upon GHS-R1a antagonist treatment in rats (Egecioglu et al. 2010). Moreover, recent studies using conditioned place preference (CPP) in rodents, demonstrated that increases in ghrelin, following peripheral administration or caloric restriction, enhances the CPP response for high-fat diet (HFD) but not chow (Egecioglu et al. 2010; Perello et al. 2010; Disse et al. 2011). Moreover, the CPP response was also shown to be dependent on the GHS-R1a receptor in the VTA, as the enhanced time spent in an environment previously paired with a palatable food reward, was not observed in GHS-R1a knockout mice (Chuang and Zigman 2010; Perello et al. 2010; Disse et al. 2011), following chemical VTA lesions or after GHS-R1a blockade in rats (Egecioglu et al. 2010). These studies clearly demonstrate that ghrelin enhances the motivation to obtain rewards in rodents in the form of palatable foods (for review see Skibicka and Dickson 2011; Perello and Zigman 2012; Schellekens et al. 2012, 2013b). In addition, operant conditioning paradigms have been used to assess the impact of ghrelin on the motivational aspects of food-associated reward (Perello et al. 2010; Skibicka and Dickson 2011, 2012; Finger et al. 2012). These studies showed direct microinjection of ghrelin into the VTA increased free feeding of chow and elevated operant responding for palatable rewards in rodents, such as sucrose or high fat pellet (Skibicka and Dickson 2011). Moreover, a progressive ratio operant conditioning paradigm demonstrated a ghrelin-induced enhancement of incentive motivation for sucrose rewards in a satiated rat and a reduced operant responding for sugar in hungry rats to the level of a satiated rat following blockade of ghrelin signaling (Skibicka et al. 2012).

Interestingly, a recent study, demonstrated that the appetite-enhancing effects of the gastric hormone ghrelin are linked to an enhanced motivation to eat, rather than to an increase in hedonic properties of food (Overduin et al. 2012). The study showed that ghrelin administration via lateral ventricle infusions markedly increased total energy intake and motivation to eat ("wanting") as assessed by a progressive operant task for sucrose, to levels comparable to or greater than those seen following food deprivation, without changes in their patterns of licking microstructure which reflects palatability ("liking"). This strongly suggests that divergent ghrelinergic neurocircuitries are mediating food reward and these are again different from the homeostatic intake effects of ghrelin, with the VTA-NAcc projections mediating ghrelin's effect on food reward but not food intake. In addition, ghrelin's effects on HFD reward response in CPP and operant conditioning were inhibited following pharmacologic or genetic blockade of the ORX receptor, suggesting that the ghrelin-mediated increase in high-fat food reward occurs in an ORX-dependent manner (Perello et al. 2010). Taken together, this data clearly demonstrates a key role for the GHS-R1a receptor at the interface between homeostatic control and hedonic eating behaviors. The prominent role of the ghrelinergic system in the regulation of feeding gives rise to it as an effective target for the development of successful anti-obesity pharmacotherapies that not only affect satiety but also selectively modulate the rewarding properties of food and reduce the desire to eat in the absence of hunger, thereby maximizing the anti-obesity potential of pharmaceutical therapies.

The Ghrelin Axis in Obesity

Ghrelin Signaling in Obesity

The key biological role of ghrelin in the regulation of food intake and energy homeostasis (Tschop et al. 2000; Nakazato et al. 2001; Kojima et al. 2004) have lead to the establishment of the ghrelinergic system as a major target for the development of anti-obesity therapeutics (Zorrilla et al. 2006; Soares and Leite-Moreira 2008; Chollet et al. 2009; Lu et al. 2009; Schellekens et al. 2009; Nass et al. 2011; Patterson et al. 2011; Costantino 2012; Delporte 2012). However, circulating levels of total ghrelin negatively correlate with adiposity and a reduced total plasma ghrelin levels is associated with general obesity (Tschop et al. 2001b; Cummings et al. 2002b; Shiiya et al. 2002). Ghrelin secretion is enhanced during fasting, malnutrition, cachexia, and in anorexia nervosa, while in obesity ghrelin levels are typically reduced, which both reflect adaptations toward altered energy balance (Tschop et al. 2001b; Krsek et al. 2003; Inui 2004; Soriano-Guillen et al. 2004; Dostalova and Haluzik 2009; Koyama et al. 2010; Yi et al. 2011; Atalayer et al. 2013). This creates an interesting conundrum whereby the usefulness of anti-obesity therapeutics targeting the ghrelinergic system comes into question and is not initially obvious since a reduction of ghrelin signaling under conditions of obesity might not have a major impact on body-weight per se. Nevertheless, while low circulating ghrelin levels are linked to obesity, visceral adipose tissue is suggested to be more sensitive to these low levels compared to subcutaneous adipose tissue, indicating that circulating ghrelin would continue to promote lipid deposition in the visceral fat depots in conditions of obesity (Kola et al. 2005). This may potentially be due to a decreased expression of GHS-R1a receptors in omental adipose tissue and higher circulating levels of the active acyl ghrelin isoform compared to des-acyl ghrelin in obesity (Rodriguez et al. 2009). Indeed, increased circulating concentrations of acylated ghrelin and decreased des-acyl ghrelin levels were found in individuals with obesity and obesity-associated type 2 diabetes, directly correlating to BMI and waist circumference. In addition, decreased protein expression levels of the GHS-R1a receptor in omental adipose tissue was shown for obese individuals (Rodriguez et al. 2009). Noteworthy, an enhanced acylated ghrelin/des-acyl ghrelin ratio in obesity is still considered to be a controversial finding as this has not been consistently demonstrated across human clinical studies (Marzullo et al. 2004).

In contrast to general obesity, high plasma ghrelin levels are observed in individuals with Prader-Willi syndrome associated obesity, which is suggested to be linked to a blunted age-related decline in total ghrelin levels (Cummings et al. 2002a; DelParigi et al. 2002). This may suggest that anti-obesity therapeutics, which directly antagonize the ghrelinergic signaling system, are particularly relevant in individuals with Prader-Willi associated obesity. Prader-Willi syndrome patients may therefore represent logical first-line candidates for testing the weight-reducing effects of ghrelin-blocking agents (Cummings et al. 2002a; Haqq et al. 2008; Schellekens et al. 2009). Moreover, obese humans displayed a much reduced postprandial suppression of ghrelin, which may reinforce obesity (le Roux et al. 2005). Interestingly, the nocturnal plasma ghrelin increase observed in healthy individuals is also blunted in obese individuals, reinforcing the aberrant ghrelin circulation in obesity (Yildiz et al. 2004). A large body of evidence demonstrates that inhibition of the ghrelin/GHS-R pathway results in reduced food intake and this may consequently lead to reductions in body weight and adiposity.

The importance of ghrelin signaling in obesity has been investigated in several rodent studies modulating gene expression of either ghrelin or its receptor, GHS-R1a (Wortley et al. 2005; Zigman et al. 2005; Shrestha et al. 2009). Initial studies using knockout mouse models of ghrelin or GHS-R1a failed to demonstrate alterations in normal food intake behavior, growth rate, or body composition suggesting that the ghrelin is not the dominant and critical regulator of appetite, despite ghrelin's potent orexigenic effects (Sun et al. 2003, 2008; Wortley et al. 2004). However, a study using RNA interference demonstrated that knockdown of GHS-R1a in the PVN of the hypothalamus did not affect daily food intake but significantly reduced body weight and blood ghrelin levels (Shrestha et al. 2009). In addition, a different study demonstrated that ghrelin knockout mice are protected from HFD induced weight gain (Wortley et al. 2004). In this study, early exposure of mice to HFD at 6 weeks of age was associated with reduced weight gain, decreased adiposity, increased loco-motor activity, and increased energy expenditure as the animals aged. Moreover, the ghrelin knockout mice in this study similarly increased fat utilization when fed a HFD (Wortley et al. 2004). The reduced accumulation of body weight and adiposity in GHS-R1a null mice on a HFD compared to control animals is in line with this finding (Zigman et al. 2005). Thus, in conclusion, ghrelin deficient mice have been found to be resistant to HFD-induced obesity presumably through the preference of utilizing fat as an energy substrate.

It is likely that in the above studies, compensatory mechanisms controlling appetite and satiety have played a role in the blunted effects on food intake in the ghrelinergic knockdown rodent models. Previously, similar minimal effects on food intake were observed when NPY or AgRP were knocked down in mice (Zigman et al. 2005), which is in line with the finding that single knockout models result in normal animals with normal food intake behaviors. Interestingly, a significant decreased body weight independent of chow intake was observed in double knockout studies, in which both the ghrelin hormone and the GHS-R1a receptor were deleted (Pflueger et al. 2008). In addition, ghrelin/GHS-R1a double knockouts displayed an increased energy expenditure and motor activity, reinforcing the evolutionary determined role of ghrelin signaling in accumulating and preserving fat reserves in times of abundance to protect against poor nutritional availability during famine.

Noteworthy, the blunted orexigenic effects of ghrelin in mice with diet-induced obesity (DIO) following a chronic HFD, suggest that obesity is associated with a ghrelin resistance suppressing the neuroendocrine ghrelin axis to limit further food intake (Perreault et al. 2004). In this study, it was demonstrated that peripheral ghrelin does not induce food intake in obese mice. However, the insensitivity of obese mice to ghrelin was again improved upon weight loss and suggests that ghrelin inhibition could prevent rebound weight gain (Perreault et al. 2004). Ghrelin resistance in mice following DIO was demonstrated to be centrally mediated in NPY/AgRP neurons of the ARC and was associated with a decreased expression and reduced ghrelin-induced secretion of NPY and AgRP (Briggs et al. 2010). The reduction in NPY/AgRP responsiveness to plasma ghrelin following DIO may function to limit further food intake. Both peripheral and centrally administered ghrelin were unable to induce food intake, reinforcing that the neuroendocrine ghrelin signaling system was suppressed in DIO mice.

Both acylated and total plasma ghrelin levels were decreased in the HFD-fed mice, and ghrelin and GOAT mRNA expression was decreased in the stomach in parallel to a decreased hypothalamic expression of GHS-R1a receptor and a lower arcuate Fos immunoreactivity. Downstream NPY/agRP neuronal targets were intact as injection of NPY intracerebroventricularly increased food intake indicating that defective NPY/AgRP function is a primary cause of ghrelin resistance. Ghrelin resistance in DIO mice was not confined to the NPY/AgRP neurons, because ghrelin did not also stimulate growth hormone secretion in the obese mice (Briggs et al. 2010). Interestingly, reduced motivation to obtain a food reward in DIO mice, as measured by a decreased operant responding, was also shown (Finger et al. 2012). Compared to lean mice, obese mice were also found to be insensitive to administration of the GHS-R antagonist (D-Lys3)-GHRP-6, which was correlated to a decreased mRNA expression of the GHS-R1a receptor in the hypothalamus and NAcc of mice on HFD. This data suggest an obesity-associated ghrelin resistance in reward-associated behaviors of food intake as well as blunted orexigenic effects in the homeostatic regulation. Noteworthy, ghrelin has several functions in the brain aside from appetite control, including mood regulation, neuroprotection, and cognitive function. This suggests that central ghrelin resistance may be involved in obesity-related cognitive decline. Thus, restoring ghrelin sensitivity may also provide therapeutic potential in maintaining healthy aging (Briggs et al. 2010).

Nevertheless, using the same mouse model of obesity (i.e., DIO) it was demonstrated that subsequent diet-induced weight loss restores NPY/AgRP neuronal responsiveness to ghrelin, which may potentially explain rebound weight gain following calorie-restricted (CR) weight loss. The dietary intervention of calorie restriction as well as change to a regular chow diet normalized body weight, glucose tolerance, plasma insulin, and total plasma ghrelin levels and ghrelin sensitivity was restored which corresponded to increases in hypothalamic NPY and AgRP mRNA expression. In addition, DIO ghrelin knockout mice exhibit reduced body weight regain after CR weight loss compared to ghrelin wild-type mice, reinforcing that ghrelin mediates the rebound weight gain following diet-induced weight loss (Briggs et al. 2013). In addition, ghrelin resistance has not yet been demonstrated in humans as both obese as well as lean subjects were responsive to intravenous administration of ghrelin, demonstrating an increased intake of food as well as an enhanced palatability of food (Druce et al. 2005).

Evolutionary speaking, the human body is hardwired with a strong drive to eat and to store excess calories as body fat (Wells 2009). Thus, it is likely that chronic long-term DIO creates a higher body weight set-point and that diet-induced weight loss provokes the brain to protect the new higher set-point, which undermine the body's physiological adjustments in chronic obesity to the previous homeostatic set-point at healthy weight. Thus, in diet-induced weight loss, the CNS and periphery respond as to a state of starvation and attempt to counteract further weight loss and try to return to the new homeostatic set-point of obesity, resulting in increased peripheral ghrelin concentrations and a restored functioning of ghrelin-responsive neuronal population in the hypothalamic ARC (Davies et al. 2009; Wells 2009). This is where specific ghrelin-targeting pharmacotherapeutics could have significant benefits and they may potentially contribute to sustained weight loss via a rewiring of the brain to the natural healthy weight set-point.

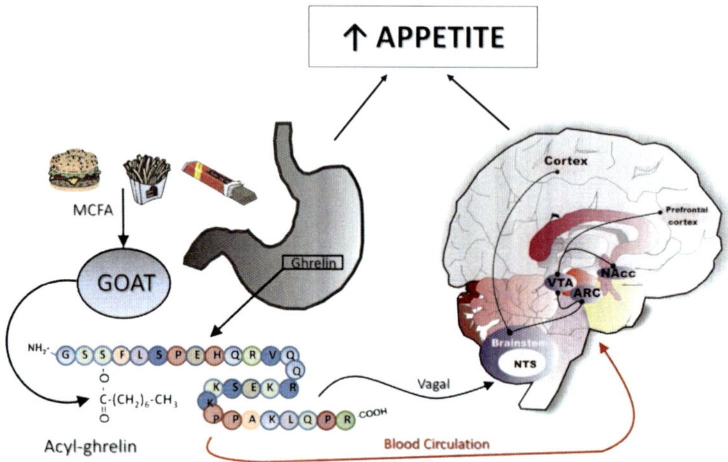

Fig. 3 *Acylation and activation of ghrelin.* The orexigenic hormone ghrelin is activated following addition of a fatty acid side chain on the third serine residue by the enzyme GOAT. Ghrelin acylation, is elevated depending on specific dietary lipids and MCFA as acylation substrates. Mature acylated ghrelin is secreted from the stomach and enters the brain via the vagus nerve or via the blood circulation after crossing the BBB. Here, ghrelin interacts with GHS-R1a receptors expressed on neurons located in several brain areas, including (but not limited to) the NTS, ARC, VTA and NAcc. Abbreviations: *ARC* arcuate nucleus of the hypothalamus; *BBB* blood brain barrier; *GOAT* ghrelin O-acyltransferase enzyme; *MCFA* medium-chain fatty acids; *NAcc* nucleus accumbens; *NTS* nucleus of the solitary tract (nucleus tractus solitarii); *VTA* ventral tegmental area (adapted from Schellekens et al. 2012)

Collectively, while ghrelin resistance in obesity again questions the potential of antagonists targeting the GHS-R1a receptor, this data also shows that the GHS-R1a antagonist can beneficially impact on the neuroadaptation following weight loss, which significantly contributes to rebound weight gain.

Ghrelin Isoforms and GHS-R Subtypes in Obesity

The mature ghrelin peptide results from enzymatic cleavage from preproghrelin, which is activated following n-octanoylation on its 3rd serine residue by GOAT (Gualillo et al. 2008; Gutierrez et al. 2008; Yang et al. 2008a, b) (Fig. 3). Interestingly, acylation of the mature ghrelin peptide is enhanced following ingestion of medium-chain fatty acids and medium-chain triacylglycerides, reinforcing the important role for endogenous ghrelin in the metabolic adaptation to nutrient availability (Nishi et al. 2005). The enhanced levels of circulating active acyl-ghrelin upon ingestion of medium-chain fatty acids may contribute to the development of DIO. Alternative modifications at the 3rd serine residue, including decanoyl, palmitoyl, benzoyl, or adamantly groups, are also capable of yielding active ghrelin (Nass et al. 2011). Indeed, circulating ghrelin is heterogenous and a variety of other natural bioactive molecules following alternative splicing, post-translational modification or produced independent of preproghrelin have been described (Hosoda et al. 2000; Tanaka et al. 2001; Funahashi et al. 2003; Jeffery et al. 2005; Kineman et al. 2007; Seim et al. 2007; Soares and Leite-Moreira 2008; Rediger et al. 2011). Processing of preproghrelin can result in the major variant 1–28 ghrelin as well as 1–27 ghrelin, which is also active following acylation of the hydroxyl group of the 3rd serine residue (Hosoda et al. 2003). In a study by Ohgusu et al., it was demonstrated that GOAT has a preference for n-hexanoyl-CoA over n-octanoyl-CoA as acyl donor, reinforcing the existence of alternatively acylated ghrelin (Ohgusu et al. 2009). It has also been shown that the N-terminal residues of ghrelin constitute the active core and are necessary for GHS-R1a binding and activation (Bednarek et al. 2000; Matsumoto et al. 2001b; Ohgusu et al. 2009).

An additional ghrelin variant, des-Gln(14)-ghrelin, results due to a variation in an intron splice junction and also contains an n-octanoyl modification on the 3rd serine residue (Hosoda et al. 2000). Des-Gln(14)-ghrelin was identified from rat stomach and shown to stimulate growth hormone release with similar potency and efficacy compared to a full-length acyl-ghrelin. Moreover, a study which isolated human ghrelin from the stomach also revealed several other ghrelin-derived molecules, including a variation in proghrelin protease cleavage sites yielding desArg(28)-ghrelin, which lacks the C-terminal Arg(28), and a series of differentially acylated ghrelin molecules, including nonacylated, octanoylated (C8:0), decanoylated (C10:0), which may constitute a double bond (C10:1) (Hosoda et al. 2003). The existence of ghrelin isoforms with different acyl groups has lead to the idea that replacement of the octanoyl ester on the ghrelin peptide by more stable ether or thioether bonds may be potentially advantageous for the generation of pharmaceuticals with longer stability (Matsumoto et al. 2001a). In addition, other ghrelin variants have been described, which include an exon-4 variant C-peptide and an In1-ghrelin variant (Jeffery et al. 2005; Kineman et al. 2007; Seim et al. 2007). Furthermore, several polymorphisms of the ghrelin gene have been identified and found to be associated with an increased risk to develop obesity (Ukkola et al. 2001; Hinney et al. 2002; Korbonits et al. 2002; Vivenza et al. 2004; Bing et al. 2005; Larsen et al. 2005).

The unacylated ghrelin isoform, des-acyl ghrelin, represents more than 90 % of human plasma ghrelin immunoreactivity and is thus the most abundant isoform in the blood circulation (Patterson et al. 2005). However, des-acyl ghrelin is unable to bind or activate the GHS-R1a receptor, which may suggest the existence of additional, yet unidentified, GHS-R subtypes (Camina 2006). Indeed, des-acyl ghrelin has been shown to modulate food intake via a yet to be identified receptor (Inhoff et al. 2008; Stengel et al. 2010). In addition, ghrelin and des-acyl ghrelin have been shown to exert some opposing biological actions (Soares and Leite-Moreira 2008). One well-characterized GHS-R subtype, already mentioned, is the GHS-R1b receptor isoform. The GHS-R1b truncated isoform may represent an interesting pharmacological target as it forms a heterodimer with the GHS-R1a receptor, attenuating its signaling (Chan and Cheng 2004; Leung et al. 2007; Schellekens et al. 2013b). Moreover, ghrelin-induced calcium mobilization has been suggested to also occur via the Gs-cAMP-PKA pathway in NPY neurons compared to the Gq11-PLC in somatotropic cells, which may also suggest different receptor subtypes (Caminos et al. 2005; Kohno et al. 2003). Receptor binding studies with radiolabeled (^{125}I)-ghrelin demonstrated binding to a functional receptor in chondrocytes, while the presence of GHS-R1a could not be demonstrated, again suggesting the existence of specific receptors different from the 1A isotype (Caminos et al. 2005). Moreover, studies using a radiolabelled GHS tracer demonstrated the presence of specific binding sites in breast carcinomas in the absence of the GHS-R1a receptor, supporting the existence of several receptors (Cassoni et al. 2001). In addition, unlabeled GHS such as hexarelin, Tyr-Ala-hexarelin, human ghrelin, and MK-0677 as well as by desoctanoyl-ghrelin and hexarelin derivative EP-80317 were all able to displace the radiolabeled GHS tracer and significantly inhibit cell proliferation at concentrations close to their binding affinity. Certain mutations in the ghrelin receptor including 4 different point mutation, I134T, V160M, A204E, and F279L, have been identified that are linked to an altered metabolism and mainly affect the constitutive activity of the GHS-R1a receptor (Liu et al. 2007). Finally, it has been shown that certain polymorphisms in the GHS-R promoter have the ability to modify GHS-R gene expression which can induce changes in body weight (Mager et al. 2008). Further studies are warranted to identify the additional GHS-R receptor subtypes, which will significantly contribute to the current knowledge on mechanisms of ghrelinergic signaling and hence may lead to better pharmacological strategies to target obesity.

GHS-R Heterodimers in Obesity

G-protein-coupled receptors (GPCRs), like the GHS-R1a receptor, were initially thought to exist and function exclusively as monomeric units. However, receptor oligomerization, whereby receptors of the same and different families combine to generate homo- or heterodimers or other multimeric complexes is becoming increasingly accepted as a fundamental process in receptor signaling (Kaupmann et al. 1998; George et al. 2002; Kent et al. 2007; Luttrell 2008; Panetta and Greenwood 2008; Smith and Milligan 2010; Teitler and Klein 2012). Indeed, many GPCR family members have a natural tendency to form oligomers upon co-expression (Salim et al. 2002) and GHS-R1a receptor heterodimerization might in fact be a common feature fine-tuning ghrelin signaling in obesity (for review see Schellekens et al. 2013b) (Fig. 4). Higher order complex systems have been suggested to exhibit unique pharmacological, biochemical, and functional characteristics such as specific signaling cascades, altered internalization, and changes in recycling properties (Hebert and Bouvier 1998; Terrillon and Bouvier 2004). Evidence is demonstrating the GHS-R1a receptor to form homodimers as well as

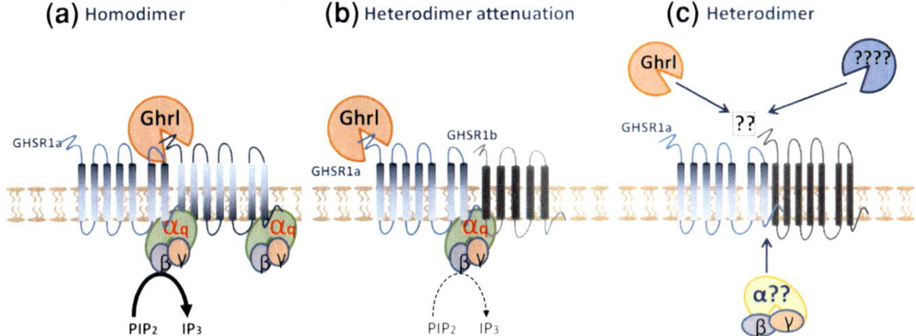

Fig. 4 *Dimerization of the ghrelin receptor.* In the homodimeric model of the ghrelin receptor binding of ghrelin occurs only in one subunit (**a**). Downstream signaling is attenuated when the GHS-R1a receptor forms a heterodimer with the truncatesd GHS-R1b isoform (**b**). Signal transduction following promiscuous heterodimerization of the GHS-R1a receptor with other GPCRs including the D_1, D_2, MC_3, GHRH, and 5-HT_{2C} receptor is dependent on the GPCR partner (**c**) (adapted from Schellekens et al. 2009)

to dimerize with other GPCRs, forming heterodimers (for review see Schellekens et al. 2013b).

The homodimeric GHS-R model was presented by Holst and colleagues, demonstrating that in the presence of the endogenous GHS-R1a agonist ghrelin, coadministration of a nonendogenous agonist can act as a neutral (MK-677), positive (L-692,429), or negative (GHRP-6) modulator of ghrelin function (Holst et al. 2005). In the study by Holst et al. (2005), heterodimerization was also suggested for the GHS-R receptor and the single transmembrane protein CD36, which binds GHRP-6 and this could explain the allosteric negative co-operative effect of GHRP-6 on ghrelin signaling (Holst et al. 2005). This same homodimeric model was also used to explain the potentiated ghrelin response when growth hormone-releasing hormone (GHRH) is also bound to the GHS-R1a receptor (Casanueva et al. 2008). The presence of GHRH did not compete for binding of ghrelin but was shown to increase the GHS-R1a-mediated calcium influx in a dose dependent-fashion and to enhance the binding capacity of ghrelin in showing a positive binding cooperativity. It was suggested that GHRH interacts with the orthosteric ghrelin binding site in absence of ghrelin and occupies the allosteric site, acting as a co-agonist in presence of endogenous ghrelin, allowing two ghrelin molecules to bind simultaneously in the two subunits of the homodimer and to increase affinity of ghrelin for the GHS-R1a receptor (Casanueva et al. 2008). Presence of GHRH increases the maximal response of ghrelin inositol phosphate turnover assays through Gq-associated signal transduction and also increases the potency in a calcium assay, but the ghrelin-mediated GH release was shown to be independent of GHRH (Takaya et al. 2000). However, the synergistic interaction could also be a consequence of a direct interaction between the GHS-R and the GHRH receptor, which is reinforced by the observed potentiation of GHRH-mediated cAMP production upon co-expression of the GHS-R1a receptor (Cunha and Mayo 2002). In addition, as previously mentioned, the GHS-R1a receptor also forms a heterodimer with its truncated splice variant, the GHS-R1b receptor, which attenuates the receptor pair in the nucleus decreasing ghrelin responsiveness, suggesting the GHS-R1b receptor to act as a dominant-negative mutant of the full-length GHS-R1a receptor (Chan and Cheng 2004; Leung et al. 2007; Muccioli et al. 2007). Homodimers of the GHS-R1a as well as heterodimers between the GHS-R1a and GHS-R1b receptor were detected using bioluminescence resonance energy transfer and co-immunoprecipitation when both receptors were expressed in human embryonic kidney (Hek) cells (Leung et al. 2007). Moreover, cell surface expression of GHS-R1a was decreased with increasing expression of GHS-R1b and a decrease of ligand-independent constitutive GHS-R1a receptor activity was also observed.

Accumulating evidence supports heterodimerization of the GHS-R1a receptor and the dopamine D_1 receptor, leading to enhanced dopamine signaling (Jiang et al. 2006). In this study, a ghrelin-mediated potentiation of dopamine-induced c-AMP accumulation in Hek cells was shown in a GHS-R-dependent manner. Interestingly, this cAMP amplification suggests a switch in G-protein coupling from Gq to Gs-mediated signaling of the GHS-R, which was previously described for the GHS-R1a receptor expressed in neuronal NPY cells of the ARC (Kohno et al. 2003). However, this ghrelin-mediated increase in cAMP may not be due to coupling of GHS-R to Gs but to ghrelin-mediated activation of another receptor subtype (Caminos et al. 2005). Nevertheless, recent data is in support of a GHS-R1a/D_1 heterodimer, as it demonstrated that co-expression of the D_1 receptor was able to attenuate GHS-R1a-mediated signaling in Hek cells (Schellekens et al. 2013c). In the same study, agonist-mediated co-internalization of the GHS-R1a/D_1 receptor pair was demonstrated, which is also in support of GHS-R1/D_1 heterodimerization. In addition, recent evidence demonstrates a dimer between the GHS-R1a and the dopamine D_2 receptor in hypothalamic neurons (Kern et al. 2012).

A subset of neurons was identified to co-express both the GHS-R1a and D_2 receptor and dimerization was shown using fluorescence energy transfer (FRET). In addition, the GHS-R1a/D_2 heterodimer was shown to allosterically modify D_2-mediated signaling, which was blocked by GHS-R1a antagonism. Interestingly, the ghrelin system has the ability to enhance preference for palatable foods rich in sugar or fat as well as to alter the rewarding value of food via modulation of the dopaminergic system. This may implicate a potential involvement of GHS-R1a receptor dimerization in hedonic appetite signaling and rewarding aspects of food, independent of homeostatic regulation of food intake, and warrants further investigation. Heterodimerization of other hypothalamic GPCRs involved in appetite signaling was also demonstrated (Rediger et al. 2009). A dimer interaction between MC_4 and GPR7 and between MC_3 and GHS-R1a was demonstrated using ELISA and FRET approaches (Rediger et al. 2009). In addition, dimerization of the GHS-R1a receptor with the MC_3 receptor was shown to attenuate the dimer pair in the cytosol (Schellekens et al. 2013c). Finally, a novel heterodimer between the GHS-R1a receptor and the 5-HT_{2C} receptor was recently identified (Schellekens et al. 2013c). Dimerization of the GHS-R1a receptor with the 5-HT_{2C} receptor was shown to significantly reduce ghrelin-mediated calcium influx, which was completely restored following pharmacological blockade of the 5-HT$_{2C}$ receptor. In conclusion, heterodimerization of the GHS-R1a receptor is likely to confer unique pharmacological and functional properties to the receptor, including differential affinity for specific peptide ligands and coupling to novel G-proteins mediating downstream signaling, depending on each specific dimer partner. Together, the promiscuous dimerization of the GHS-R1a receptor indicates a potential novel mechanism for fine-tuning GHS-R1a receptor-mediated activity, which significant implications for the development of future pharmacotherapeutics targeting of the GHS-R1a receptor in the homeostatic regulation of energy balance and in hedonic appetite signaling. The specific targeting of GHS-R1a heterodimers presents the pharmaceutical industry with novel strategies to modulate appetite and food intake in metabolic eating disorders including obesity.

Conclusion and Future Perspectives

The rapidly increasing incidence of obesity in modern day society is a growing concern as it is one of the leading causes of illness and mortality. However, appropriate effective pharmacological interventions to reduce body weight and to alleviate comorbidities associated with the metabolic syndrome are still lacking, highlighting the need for novel treatment strategies. Ghrelin and its receptor, the growth hormone secretagogue (GHS-R1a) receptor, have both been strongly conserved during evolution, reinforcing their fundamentally important role in biology (Palyha et al. 2000). The ghrelin axis has a particular major biological function in the multi-process neurocircuitries of feeding. Ghrelin is the only peripheral derived orexigenic hormone and regulates energy homeostasis as well as hedonic food intake (Tschop et al. 2000; Nakazato et al. 2001; Kojima et al. 2004). The gastric hormone ghrelin plays a key role in the sensation of hunger and meal initiation due to its orexigenic effects, and circulating plasma ghrelin levels are correlated with nutrient availability, with increased levels just before a meal followed by postprandial decrease (Tschop et al. 2000; Cummings et al. 2001). These circulating ghrelin levels are dysregulated in metabolic eating disorders, including obesity. The GHS-R1a receptor has, therefore, been a pharmacological target in the development of anti-obesity pharmaceuticals in the last decade (for review see Soares et al. 2008; Chollet et al. 2009; Schellekens et al. 2009; Patterson et al. 2011; Delporte 2012). Indeed, immunoneutralization of the ghrelin peptide or pharmacological blockade of the GHS-R1a receptor using GHS-R1a specific antagonists has been shown to decrease food intake and reduce adiposity in normal and in leptin deficient mice (Nakazato et al. 2001; Asakawa et al. 2003; Zorrilla et al. 2006). The peptide GHS-R1a antagonist [D-Lys^3]-GHRP-6 decreased food intake in lean and obese mice and reduced weight gain (Asakawa et al. 2003; Beck et al. 2004; Finger et al. 2011). In addition, piperidine-substituted quinazolinone derivatives were identified as a new class of small-molecule GHS-R1a antagonists, suppressing food intake and reducing body weight as well as stimulating glucose-dependent insulin secretion (Rudolph et al. 2007). The piperidine-substituted quinazoline derivative, YIL-781, acted as a potent GHS-R1a antagonist, stimulated weight loss by reducing food intake and also improved glucose-stimulated insulin secretion (Esler et al. 2007). Interestingly, some GHS-R1a analogs have been shown to behave as GHS-R1a antagonist and are therefore also considered as preclinical compounds to treat weight disorders, including obesity. These include TZP-301 developed by Tranzyme Pharma, EX-1350, from Elixir Pharmaceuticals (Depoortere 2009), the triazole derivatives JMV2866 and JMV2844 (Demange et al. 2007; Moulin et al. 2008a, b). More recently, pipera-zine-bisamide analogs were discovered as partial agonists of the GHS-R1a receptor and optimized for potency and converted into antagonists through structure-activity relationship (SAR) studies (Yu et al. 2010). In addition, inverse GHS-R1a agonists, such as [D-Arg1, D-Phe5, D-Trp7,9, Leu11] substance P, which decrease the high constitutive activity of the GHS-R1a receptor are also being investigated for the treatment of obesity (Holst et al. 2003; Holst and Schwartz 2004; Holliday et al. 2007). Many additional GHS-R ligands and strat-egies are being developed as pharmacological tools to inhibit GHS-R1a activity, which are extensively reviewed elsewhere (Schellekens et al. 2009). Despite this, no successful anti-obesity pharmacological treatments targeting the GHS-R1a receptor are currently on the market as long-term studies in animal models and humans are lacking. Further studies are needed to evaluate the beneficial properties and usefulness of GHS-R1a ligands in the treatment of obesity. Nevertheless, it is still believed that inhibition of ghrelin signaling via GHS-R1a receptor antagonists represents an attractive target for the future pharmacological treatment of obesity, in particular Prader-Willi syndrome. In addition, ghrelin has been recognized to also regulate glucose-induced insulin secretion, which suggests that GHS-R1a antagonists may also have potential value in the treatment of diabetes. Moreover, the existence of ghrelin isoforms with different acyl groups has lead to the idea that replacement of the octanoyl ester on the ghrelin peptide by more stable ether or thioether bonds may be potentially advantageous for the generation of pharma-ceuticals with longer stability (Matsumoto et al. 2001a). In addition, evidence is accumulating strongly suggesting the existence of an additional GHS-R1a receptor isoform, which may explain the effects of des-acyl ghrelin on food intake (Inhoff et al. 2008; Stengel et al. 2010). Identification of this isoform will enhance our understanding and is likely to represent an additional target in the development of anti-obesity therapeutics.

A potential novel strategy to curb appetite in obesity could be the targeting of specific GHS-R1a heterodimers, which may also have the added benefit of an improved specificity with less side effects (Schellekens et al. 2013b). Dimerization of the GHS-R1a receptor can function to fine-tune ghrelin signaling and modu-lation of the GHS-R1a receptor is poised to also affect D_1 and D_2 signaling when complexed in dimers (Jiang et al. 2006; Kern et al. 2012; Schellekens et al. 2013c). This may impact on ghrelin-mediated reward-driven hedonic eating behavior and warrants further investigation. The neuronal circuits in the hypothalamus and brainstem mediate homeostatic control, while the mesolimbic dopaminergic circuitry mediates the eating behavior associated with hedonic feeding and food palatability. The GHS-R1a[®] is ubiquitously expressed in both neurocircuitry, including within hypothalamic neurons and in cortical areas as well as the VTA, NAcc, hippocampus, and amygdala, reinforcing its role at the interface between homeostatic control and neurobiological circuits involved in reward and motivational aspects of food (for review see Dickson et al. 2011; Egecioglu et al. 2011; Skibicka and Dickson 2011). Thus, the direct effects of ghrelin on the hypothalamic nuclei regulate energy homeostasis, while direct effects of ghrelin on VTA and NAcc affect reward and incentive motivational aspects of food intake and this may involve signaling via heterodimers between the GHS-R1a receptor and dopamine receptors. Furthermore, heterodimers have been shown between the GHS-R1a receptor and the GHS-R1b, GHRH, MC3, and 5-HT$_{2C}$ receptors (Casanueva et al. 2008; Rediger et al. 2009; Schellekens et al. 2013c). The implications of GHS-R1a receptor heterodimerization fundamentally changes our current knowledge on ghrelin signaling, which will have significant consequences for drug development and screening. The enhanced pharmacological diversity of the GHS-R1a receptor opens up new avenues for the development of potential novel anti-obesity therapeutics targeting the GHS-R1a receptor with increased selectivity (Panetta and Greenwood 2008; Rozenfeld and Devi 2010, 2011). In conclusion, the GHS-R1a receptor represents an excellent pharmacological target with therapeutic potential for the development of novel future treatment strategies to combat obesity through targeting GHS-R1a receptor dimers, including receptors, which are involved in both the homeostatic as hedonic control of food intake. Deciphering the downstream signaling mechanisms following dimerization of the GHS-R1a receptor in vivo, would contribute significantly to the knowledge needed by the pharmaceutical industry to develop appropriate therapeutic strategies modulating homeostatic food intake as well as to modify the incentive salience and rewarding properties of food, which are both mediated via the GHS-R1a receptor.

Acknowledgements The work was supported by Enterprise Ireland under Grant Number CC20080001. JFC and TGD are also supported in part by Science Foundation Ireland (SFI) in the form of a center grant (Alimentary Pharmabiotic Center) through the Irish Government's National Development Plan. The authors and their work were supported by SFI (grant no.s 02/CE/B124 and 07/CE/B1368). JFC is funded by European Community's Seventh Framework Program; Grant Number: FP7/2007-2013, Grant Agreement 201714.

References

Abizaid A (2009) Ghrelin and dopamine: new insights on the peripheral regulation of appetite. J Neuroendocrinol 21(9):787–793

Abizaid A, Liu ZW, Andrews ZB, Shanabrough M, Borok E, Elsworth JD, Roth RH, Sleeman MW, Picciotto MR, Tschop MH, Gao XB, Horvath TL (2006) Ghrelin modulates the activity and synaptic input organization of midbrain dopamine neurons while promoting appetite. J Clin Invest 116(12):3229–3239

Adan RA, Tiesjema B, Hillebrand JJ, La Fleur SE, Kas MJ, Krom M de (2006) The MC4 receptor and control of appetite. Br J Pharmacol 149(7): 815–27

Ahima RS, Antwi DA (2008) Brain regulation of appetite and satiety. Endocrinol Metab Clin North Am 37(4):811–823

Ahima RS, Lazar MA (2008) Adipokines and the peripheral and neural control of energy balance. Mol Endocrinol 22(5):1023–1031

Andrews ZB (2011) Central mechanisms involved in the orexigenic actions of ghrelin. Peptides 32(11):2248–2255

Andrews ZB, Liu ZW, Walllingford N, Erion DM, Borok E, Friedman JM, Tschop MH, Shanabrough M, Cline G, Shulman GI, Coppola A, Gao XB, Horvath TL, Diano S (2008) UCP2 mediates ghrelin's action on NPY/AgRP neurons by lowering free radicals. Nature 454(7206):846–851

Arnold M, Mura A, Langhans W, Geary N (2006) Gut vagal afferents are not necessary for the eating-stimulatory effect of intraperitoneally injected ghrelin in the rat. J Neurosci 26(43):11052–11060

Asakawa A, Inui A, Kaga T, Katsuura G, Fujimiya M, Fujino MA, Kasuga M (2003) Antagonism of ghrelin receptor reduces food intake and body weight gain in mice. Gut 52(7):947–952

Atalayer D, Gibson C, Konopacka A, Geliebter A (2013) Ghrelin and eating disorders. Prog Neuropsychopharmacol Biol Psychiatry 40:70–82

Banks WA, Tschop M, Robinson SM, Heiman ML (2002) Extent and direction of ghrelin transport across the blood-brain barrier is determined by its unique primary structure. J Pharmacol Exp Ther 302(2):822–827

Banks WA, Burney BO, Robinson SM (2008) Effects of triglycerides, obesity, and starvation on ghrelin transport across the blood-brain barrier. Peptides 29(11):2061–2065

Bassareo V, Di Chiara G (1999) Differential responsiveness of dopamine transmission to food-stimuli in nucleus accumbens shell/core compartments. Neuroscience 89(3):637–641

Beck B, Richy S, Stricker-Krongrad A (2004) Feeding response to ghrelin agonist and antagonist in lean and obese Zucker rats. Life Sci 76(4):473–478

Bednarek MA, Feighner SD, Pong SS, McKee KK, Hreniuk DL, Silva MV, Warren VA, Howard AD, Van Der Ploeg LH, Heck JV (2000) Structure-function studies on the new growth hormone-releasing peptide, ghrelin: minimal sequence of ghrelin necessary for activation of growth hormone secretagogue receptor 1a. J Med Chem 43(23):4370–4376

Bing C, Ambye L, Fenger M, Jorgensen T, Borch-Johnsen K, Madsbad S, Urhammer SA (2005) Large-scale studies of the Leu72Met polymorphism of the ghrelin gene in relation to the metabolic syndrome and associated quantitative traits. Diabet Med 22(9):1157–1160

Blevins JE, Baskin DG (2010) Hypothalamic-brainstem circuits controlling eating. Forum Nutr 63:133–140

Bloom SR, Kuhajda FP, Laher I, Pi-Sunyer X, Ronnett GV, Tan TM, Weigle DS (2008) The obesity epidemic: pharmacological challenges. Mol Interv 8(2):82–98

Briggs DI, Enriori PJ, Lemus MB, Cowley MA, Andrews ZB (2010) Diet-Induced Obesity Causes Ghrelin Resistance in Arcuate NPY/AgRP Neurons. Endocrinology 151(10):4745–4755

Briggs DI, Lockie SH, Wu Q, Lemus MB, Stark R, Andrews ZB (2013) Calorie-restricted weight loss reverses high-fat diet-induced ghrelin resistance, which contributes to rebound weight gain in a ghrelin-dependent manner. Endocrinology 154(2):709–717

Camina JP (2006) Cell biology of the ghrelin receptor. J Neuroendocrinol 18(1):65–76

Caminos JE, Gualillo O, Lago F, Otero M, Blanco M, Gallego R, Garcia-Caballero T, Goldring MB, Casanueva FF, Gomez-Reino JJ, Dieguez C (2005) The endogenous growth hormone secretagogue (ghrelin) is synthesized and secreted by chondrocytes. Endocrinology 146(3):1285–1292

Casanueva FF, Camina JP, Carreira MC, Pazos Y, Varga JL, Schally AV (2008) Growth hormone-releasing hormone as an agonist of the ghrelin receptor GHS-R1a. Proc Natl Acad Sci USA 105(51):20452–20457

Castaneda TR, Tong J, Datta R, Culler M, Tschop MH (2010) Ghrelin in the regulation of body weight and metabolism. Front Neuroendocrinol 31(1):44–60

Center_for_Disease_Control (2013) "http://www.cdc.gov/obesity/data/databases.html"

Chakrabarti R (2009) Pharmacotherapy of obesity: emerging drugs and targets. Expert Opin Ther Targets 13(2):195–207

Chan CB, Cheng CH (2004) Identification and functional characterization of two alternatively spliced growth hormone secretagogue receptor transcripts from the pituitary of black seabream Acanthopagrus schlegeli. Mol Cell Endocrinol 214(1–2):81–95

Chen HY, Trumbauer ME, Chen AS, Weingarth DT, Adams JR, Frazier EG, Shen Z, Marsh DJ, Feighner SD, Guan XM, Ye Z, Nargund RP, Smith RG, Ploeg LH Van der, Howard AD, MacNeil DJ, Qian S (2004) Orexigenic action of peripheral ghrelin is mediated by neuropeptide Y and agouti-related protein. Endocrinology 145(6): 2607–2612

Cheng AY, Leiter LA (2006) Metabolic syndrome under fire: weighing in on the truth. Can J Cardiol 22(5):379–382

Chollet C, Meyer K, Beck-Sickinger AG (2009) Ghrelin–a novel generation of anti-obesity drug: design, pharmacomodulation and biological activity of ghrelin analogues. J Pept Sci 15(11):711–730

Chuang JC, Zigman JM (2010) Ghrelin's roles in stress, mood, and anxiety regulation. Int J Pept

Connolly HM, Crary JL, McGoon MD, Hensrud DD, Edwards BS, Edwards WD, Schaff HV (1997) Valvular heart disease associated with fenfluramine-phentermine. N Engl J Med 337(9):581–588

Costantino L (2012) Growth hormone secretagogue receptor antagonists. Expert Opin Ther Pat 22(6):697–700

Cowley MA, Smith RG, Diano S, Tschop M, Pronchuk N, Grove KL, Strasburger CJ, Bidlingmaier M, Esterman M, Heiman ML, Garcia-Segura LM, Nillni EA, Mendez P, Low MJ, Sotonyi P, Friedman JM, Liu H, Pinto S, Colmers WF, Cone RD, Horvath TL (2003) The distribution and mechanism of action of ghrelin in the CNS demonstrates a novel hypothalamic circuit regulating energy homeostasis. Neuron 37(4):649–661

Cummings DE, Shannon MH (2003) Roles for ghrelin in the regulation of appetite and body weight. Arch Surg 138(4):389–396

Cummings DE, Purnell JQ, Frayo RS, Schmidova K, Wisse BE, Weigle DS (2001) A preprandial rise in plasma ghrelin levels suggests a role in meal initiation in humans. Diabetes 50(8):1714–1719

Cummings DE, Clement K, Purnell JQ, Vaisse C, Foster KE, Frayo RS, Schwartz MW, Basdevant A, Weigle DS (2002a) Elevated plasma ghrelin levels in Prader Willi syndrome. Nat Med 8(7):643–644

Cummings DE, Weigle DS, Frayo RS, Breen PA, Ma MK, Dellinger EP, Purnell JQ (2002b) Plasma ghrelin levels after diet-induced weight loss or gastric bypass surgery. N Engl J Med 346(21):1623–1630

Currie PJ, Mirza A, Fuld R, Park D, Vasselli JR (2005) Ghrelin is an orexigenic and metabolic signaling peptide in the arcuate and paraventricular nuclei. Am J Physiol Regul Integr Comp Physiol 289(2):R353–R358

Date Y, Kojima M, Hosoda H, Sawaguchi A, Mondal MS, Suganuma T, Matsukura S, Kangawa K, Nakazato M (2000) Ghrelin, a novel growth hormone-releasing acylated peptide, is synthesized in a distinct endocrine cell type in the gastrointestinal tracts of rats and humans. Endocrinology 141(11):4255–4261

DelParigi A, Tschop M, Heiman ML, Salbe AD, Vozarova B, Sell SM, Bunt JC, Tataranni PA (2002) High circulating ghrelin: a potential cause for hyperphagia and obesity in prader-willi syndrome. J Clin Endocrinol Metab 87(12):5461–5464

Delporte C (2012) Recent advances in potential clinical application of ghrelin in obesity. J Obes 2012:535624

Demange L, Boeglin D, Moulin A, Mousseaux D, Ryan J, Berge G, Gagne D, Heitz A, Perrissoud D, Locatelli V, Torsello A, Galleyrand JC, Fehrentz JA, Martinez J (2007) Synthesis and pharmacological in vitro and in vivo evaluations of novel triazole derivatives as ligands of the ghrelin receptor. 1. J Med Chem 50(8):1939–1957

Depoortere I (2009) Targeting the ghrelin receptor to regulate food intake. Regul Pept 156(1–3):13–23

Derosa G, Maffioli P (2012) Anti-obesity drugs: a review about their effects and their safety. Expert Opin Drug Saf 11(3):459–471

Dickson SL, Egecioglu E, Landgren S, Skibicka KP, Engel JA, Jerlhag E (2011) The role of the central ghrelin system in reward from food and chemical drugs. Mol Cell Endocrinol 340(1):80–87

Disse E, Bussier AL, Veyrat-Durebex C, Deblon N, Pfluger PT, Tschop MH, Laville M, Rohner-Jeanrenaud F (2010) Peripheral ghrelin enhances sweet taste food consumption and preference, regardless of its caloric content. Physiol Behav 101(2):277–281

Disse E, Bussier AL, Deblon N, Pfluger PT, Tschop MH, Laville M, Rohner-Jeanrenaud F (2011) Systemic ghrelin and reward: effect of cholinergic blockade. Physiol Behav 102(5):481–484

Dostalova I, Haluzik M (2009) The role of ghrelin in the regulation of food intake in patients with obesity and anorexia nervosa. Physiol Res 58(2):159–170

Douglas A, Douglas JG, Robertson CE, Munro JF (1983) Plasma phentermine levels, weight loss and side-effects. Int J Obes 7(6):591–595

Druce MR, Wren AM, Park AJ, Milton JE, Patterson M, Frost G, Ghatei MA, Small C, Bloom SR (2005) Ghrelin increases food intake in obese as well as lean subjects. Int J Obes (Lond) 29(9):1130–1136

Egecioglu E, Jerlhag E, Salome N, Skibicka KP, Haage D, Bohlooly YM, Andersson D, Bjursell M, Perrissoud D, Engel JA, Dickson SL (2010) Ghrelin increases intake of rewarding food in rodents. Addict Biol 15(3):304–311

Egecioglu E, Skibicka KP, Hansson C, Alvarez-Crespo M, Friberg PA, Jerlhag E, Engel JA, Dickson SL (2011) Hedonic and incentive signals for body weight control. Rev Endocr Metab Disord 12(3):141–151

Eknoyan G (2008) Adolphe Quetelet (1796-1874)–the average man and indices of obesity. Nephrol Dial Transplant 23(1):47–51

Esler WP, Rudolph J, Claus TH, Tang W, Barucci N, Brown SE, Bullock W, Daly M, Decarr L, Li Y, Milardo L, Molstad D, Zhu J, Gardell SJ, Livingston JN, Sweet LJ (2007) Small-molecule ghrelin receptor antagonists improve glucose tolerance, suppress appetite, and promote weight loss. Endocrinology 148(11):5175–5185

Evans WJ, Morley JE, Argiles J, Bales C, Baracos V, Guttridge D, Jatoi A, Kalantar-Zadeh K, Lochs H, Mantovani G, Marks D, Mitch WE, Muscaritoli M, Najand A, Ponikowski P, Rossi Fanelli F, Schambelan M, Schols A, Schuster M, Thomas D, Wolfe R, Anker SD (2008) Cachexia: a new definition. Clin Nutr 27(6):793–799

Faulconbridge LF, Cummings DE, Kaplan JM, Grill HJ (2003) Hyperphagic effects of brainstem ghrelin administration. Diabetes 52(9):2260–2265

Finger BC, Schellekens H, Dinan TG, Cryan JF (2011) Is there altered sensitivity to ghrelin-receptor ligands in leptin-deficient mice?: importance of satiety state and time of day. Psychopharmacology 216(3):421–429

Finger BC, Dinan TG, Cryan JF (2012) Diet-induced obesity blunts the behavioural effects of ghrelin: studies in a mouse-progressive ratio task. Psychopharmacology 220(1):173–181

Flegal KM, Carroll MD, Kit BK, Ogden CL (2012) Prevalence of obesity and trends in the distribution of body mass index among US adults, 1999–2010. JAMA 307(5):491–497

Funahashi H, Takenoya F, Guan JL, Kageyama H, Yada T, Shioda S (2003) Hypothalamic neuronal networks and feeding-related peptides involved in the regulation of feeding. Anat Sci Int 78(3):123–138

Furness JB, Hunne B, Matsuda N, Yin L, Russo D, Kato I, Fujimiya M, Patterson M, McLeod J, Andrews ZB, Bron R (2011) Investigation of the presence of ghrelin in the central nervous system of the rat and mouse. Neuroscience 193:1–9

George SR, O'Dowd BF, Lee SP (2002) G-protein-coupled receptor oligomerization and its potential for drug discovery. Nat Rev Drug Discov 1(10):808–820

Gibbs J, Young RC, Smith GP (1973) Cholecystokinin elicits satiety in rats with open gastric fistulas. Nature 245(5424):323–325

Gualillo O, Lago F, Dieguez C (2008) Introducing GOAT: a target for obesity and anti-diabetic drugs? Trends Pharmacol Sci 29(8):398–401

Guan XM, Yu H, Palyha OC, McKee KK, Feighner SD, Sirinathsinghji DJ, Smith RG, Ploeg LH Van der, Howard AD (1997) Distribution of mRNA encoding the growth hormone secretagogue receptor in brain and peripheral tissues. Brain Res Mol Brain Res 48(1):23–29

Gutierrez JA, Solenberg PJ, Perkins DR, Willency JA, Knierman MD, Jin Z, Witcher DR, Luo S, Onyia JE, Hale JE (2008) Ghrelin octanoylation mediated by an orphan lipid transferase. Proc Natl Acad Sci U S A 105(17):6320–6325

Halford JC, Boyland EJ, Blundell JE, Kirkham TC, Harrold JA (2010) Pharmacological management of appetite expression in obesity. Nat Rev Endocrinol 6(5):255–269

Haqq AM, Grambow SC, Muehlbauer M, Newgard CB, Svetkey LP, Carrel AL, Yanovski JA, Purnell JQ, Freemark M (2008) Ghrelin concentrations in Prader-Willi syndrome (PWS) infants and children: changes during development. Clin Endocrinol (Oxf) 69(6):911–920

Hebert TE, Bouvier M (1998) Structural and functional aspects of G protein-coupled receptor oligomerization. Biochem Cell Biol 76(1):1–11

Hinney A, Hoch A, Geller F, Schafer H, Siegfried W, Goldschmidt H, Remschmidt H, Hebebrand J (2002) Ghrelin gene: identification of missense variants and a frameshift mutation in extremely obese children and adolescents and healthy normal weight students. J Clin Endocrinol Metab 87(6):2716

Hoebel BG (1985) Brain neurotransmitters in food and drug reward. Am J Clin Nutr 42(5 Suppl):1133–1150

Holliday ND, Holst B, Rodionova EA, Schwartz TW, Cox HM (2007) Importance of constitutive activity and arrestin-independent mechanisms for intracellular trafficking of the ghrelin receptor. Mol Endocrinol 21(12):3100–3112

Holst B, Schwartz TW (2004) Constitutive ghrelin receptor activity as a signaling set-point in appetite regulation. Trends Pharmacol Sci 25(3):113–117

Holst B, Cygankiewicz A, Jensen TH, Ankersen M, Schwartz TW (2003) High constitutive signaling of the ghrelin receptor–identification of a potent inverse agonist. Mol Endocrinol 17(11):2201–2210

Holst B, Brandt E, Bach A, Heding A, Schwartz TW (2005) Nonpeptide and peptide growth hormone secretagogues act both as ghrelin receptor agonist and as positive or negative allosteric modulators of ghrelin signaling. Mol Endocrinol 19(9):2400–2411

Hosoda H, Kojima M, Matsuo H, Kangawa K (2000) Purification and characterization of rat des-Gln14-Ghrelin, a second endogenous ligand for the growth hormone secretagogue receptor. J Biol Chem 275(29):21995–22000

Hosoda H, Kojima M, Mizushima T, Shimizu S, Kangawa K (2003) Structural divergence of human ghrelin. Identification of multiple ghrelin-derived molecules produced by post-translational processing. J Biol Chem 278(1):64–70

Hou Z, Miao Y, Gao L, Pan H, Zhu S (2006) Ghrelin-containing neuron in cerebral cortex and hypothalamus linked with the DVC of brainstem in rat. Regul Pept 134(2–3):126–131

Howard AD, Feighner SD, Cully DF, Arena JP, Liberator PA, Rosenblum CI, Hamelin M, Hreniuk DL, Palyha OC, Anderson J, Paress PS, Diaz C, Chou M, Liu KK, McKee KK, Pong SS, Chaung LY, Elbrecht A, Dashkevicz M, Heavens R, Rigby M, Sirinathsinghji DJ, Dean DC, Melillo DG, Patchett AA, Nargund R, Griffin PR, DeMartino JA, Gupta SK, Schaeffer JM, Smith RG, Van der Ploeg LH (1996) A receptor in pituitary and hypothalamus that functions in growth hormone release. Science 273(5277):974–977

Inhoff T, Monnikes H, Noetzel S, Stengel A, Goebel M, Dinh QT, Riedl A, Bannert N, Wisser AS, Wiedenmann B, Klapp BF, Tache Y, Kobelt P (2008) Desacyl ghrelin inhibits the orexigenic effect of peripherally injected ghrelin in rats. Peptides 29(12):2159–2168

Inui A (2004) Ghrelin, obesity and anorexia nervosa. J Pediatr 145(6):862; author reply 862–823

Jeffery PL, Duncan RP, Yeh AH, Jaskolski RA, Hammond DS, Herington AC, Chopin LK (2005) Expression of the ghrelin axis in the mouse: an exon 4-deleted mouse proghrelin variant encodes a novel C terminal peptide. Endocrinology 146(1):432–440

Jerlhag E, Egecioglu E, Dickson SL, Douhan A, Svensson L, Engel JA (2007) Ghrelin administration into tegmental areas stimulates locomotor activity and increases extracellular concentration of dopamine in the nucleus accumbens. Addict Biol 12(1):6–16

Jiang H, Betancourt L, Smith RG (2006) Ghrelin amplifies dopamine signaling by cross talk involving formation of growth hormone secretagogue receptor/dopamine receptor subtype 1 heterodimers. Mol Endocrinol 20(8):1772–1785

Kadowaki T, Yamauchi T, Kubota N (2008) The physiological and pathophysiological role of adiponectin and adiponectin receptors in the peripheral tissues and CNS. FEBS Lett 582(1):74–80

Kageyama H, Kitamura Y, Hosono T, Kintaka Y, Seki M, Takenoya F, Hori Y, Nonaka N, Arata S, Shioda S (2008) Visualization of ghrelin-producing neurons in the hypothalamic arcuate nucleus using ghrelin-EGFP transgenic mice. Regul Pept 145(1–3):116–121

Kageyama K, Akimoto K, Yamagata S, Sugiyama A, Murasawa S, Watanuki Y, Tamasawa N, Suda T (2012) Dexamethasone stimulates the expression of ghrelin and its receptor in rat hypothalamic 4B cells. Regul Pept 174(1–3):12–17

Kang JG, Park CY (2012) Anti-Obesity drugs: a review about their effects and safety. Diabetes Metab J 36(1):13–25

Kaupmann K, Malitschek B, Schuler V, Heid J, Froestl W, Beck P, Mosbacher J, Bischoff S, Kulik A, Shigemoto R, Karschin A, Bettler B (1998) GABA(B)-receptor subtypes assemble into functional heteromeric complexes. Nature 396(6712):683–687

Kaye WH, Wierenga CE, Bailer UF, Simmons AN, Bischoff-Grethe A (2013) Nothing tastes as good as skinny feels: the neurobiology of anorexia nervosa. Trends Neurosci 36(2):110–120

Kent T, McAlpine C, Sabetnia S, Presland J (2007) G-protein-coupled receptor heterodimerization: assay technologies to clinical significance. Curr Opin Drug Discov Devel 10(5):580–589

Kern A, Albarran-Zeckler R, Walsh HE, Smith RG (2012) Apo-ghrelin receptor forms heteromers with DRD2 in hypothalamic neurons and is essential for anorexigenic effects of DRD2 agonism. Neuron 73(2):317–332

Kineman RD, Gahete MD, Luque RM (2007) Identification of a mouse ghrelin gene transcript that contains intron 2 and is regulated in the pituitary and hypothalamus in response to metabolic stress. J Mol Endocrinol 38(5):511–521

Kohno D, Gao HZ, Muroya S, Kikuyama S, Yada T (2003) Ghrelin directly interacts with neuropeptide-Y-containing neurons in the rat arcuate nucleus: Ca2+ signaling via protein kinase A and N-type channel-dependent mechanisms and cross-talk with leptin and orexin. Diabetes 52(4):948–956

Kojima M, Kangawa K (2002) Ghrelin, an orexigenic signaling molecule from the gastrointestinal tract. Curr Opin Pharmacol 2(6):665–668

Kojima M, Hosoda H, Date Y, Nakazato M, Matsuo H, Kangawa K (1999) Ghrelin is a growth-hormone-releasing acylated peptide from stomach. Nature 402(6762):656–660

Kojima M, Hosoda H, Kangawa K (2001) Purification and distribution of ghrelin: the natural endogenous ligand for the growth hormone secretagogue receptor. Horm Res 56(Suppl 1):93–97

Kojima M, Hosoda H, Kangawa K (2004) Clinical endocrinology and metabolism. Ghrelin, a novel growth-hormone-releasing and appetite-stimulating peptide from stomach. Best Pract Res Clin Endocrinol Metab 18(4):517–530

Kola B, Hubina E, Tucci SA, Kirkham TC, Garcia EA, Mitchell SE, Williams LM, Hawley SA, Hardie DG, Grossman AB, Korbonits M (2005) Cannabinoids and ghrelin have both central and peripheral metabolic and cardiac effects via AMP-activated protein kinase. J Biol Chem 280(26):25196–25201

Konturek SJ, Konturek JW, Pawlik T, Brzozowski T (2004) Brain-gut axis and its role in the control of food intake. J Physiol Pharmacol 55(1 Pt 2):137–154

Korbonits M, Gueorguiev M, O'Grady E, Lecoeur C, Swan DC, Mein CA, Weill J, Grossman AB, Froguel P (2002) A variation in the ghrelin gene increases weight and decreases insulin secretion in tall, obese children. J Clin Endocrinol Metab 87(8):4005–4008

Koyama KI, Yasuhara D, Nakahara T, Harada T, Uehara M, Ushikai M, Asakawa A, Inui A (2010) Changes in acyl ghrelin, des-acyl ghrelin, and ratio of acyl ghrelin to total ghrelin with short-term refeeding in female inpatients with restricting-type anorexia nervosa. Horm Metab Res 42(8):595–598

Krsek M, Rosicka M, Papezova H, Krizova J, Kotrlikova E, Haluz'k M, Justova V, Lacinova Z, Jarkovska Z (2003) Plasma ghrelin levels and malnutrition: a comparison of two etiologies. Eat Weight Disord 8(3):207–211

Kurose Y, Iqbal J, Rao A, Murata Y, Hasegawa Y, Terashima Y, Kojima M, Kangawa K, Clarke IJ (2005) Changes in expression of the genes for the leptin receptor and the growth hormone-releasing peptide/ghrelin receptor in the hypothalamic arcuate nucleus with long-term manipulation of adiposity by dietary means. J Neuroendocrinol 17(6):331–340

Larsen LH, Gjesing AP, Sorensen TI, Hamid YH, Echwald SM, Toubro S, Black E, Astrup A, Hansen T, Pedersen O (2005) Mutation analysis of the preproghrelin gene: no association with obesity and type 2 diabetes. Clin Biochem 38(5):420–424

Lawrence CB, Snape AC, Baudoin FM, Luckman SM (2002) Acute central ghrelin and GH secretagogues induce feeding and activate brain appetite centers. Endocrinology 143(1):155–162

le Roux CW, Patterson M, Vincent RP, Hunt C, Ghatei MA, Bloom SR (2005) Postprandial plasma ghrelin is suppressed proportional to meal calorie content in normal-weight but not obese subjects. J Clin Endocrinol Metab 90(2):1068–1071

Leite-Moreira AF, Soares JB (2007) Physiological, pathological and potential therapeutic roles of ghrelin. Drug Discov Today 12(7–8):276–288

Leung PK, Chow KB, Lau PN, Chu KM, Chan CB, Cheng CH, Wise H (2007) The truncated ghrelin receptor polypeptide (GHS-R1b) acts as a dominant-negative mutant of the ghrelin receptor. Cell Signal 19(5):1011–1022

Liu G, Fortin JP, Beinborn M, Kopin AS (2007) Four missense mutations in the ghrelin receptor result in distinct pharmacological abnormalities. J Pharmacol Exp Ther 322(3):1036–1043

Lu S, Guan JL, Wang QP, Uehara K, Yamada S, Goto N, Date Y, Nakazato M, Kojima M, Kangawa K, Shioda S (2002) Immunocytochemical observation of ghrelin-containing neurons in the rat arcuate nucleus. Neurosci Lett 321(3):157–160

Lu SC, Xu J, Chinookoswong N, Liu S, Steavenson S, Gegg C, Brankow D, Lindberg R, Veniant M, Gu W (2009) An acyl-ghrelin-specific neutralizing antibody inhibits the acute ghrelin-mediated orexigenic effects in mice. Mol Pharmacol 75(4):901–907

Luttrell LM (2008) Reviews in molecular biology and biotechnology: transmembrane signaling by G protein-coupled receptors. Mol Biotechnol 39(3):239–264

Lutz TA (2006) Amylinergic control of food intake. Physiol Behav 89(4):465–471

Mager U, Degenhardt T, Pulkkinen L, Kolehmainen M, Tolppanen AM, Lindstrom J, Eriksson JG, Carlberg C, Tuomilehto J, Uusitupa M (2008) Variations in the ghrelin receptor gene associate with obesity and glucose metabolism in individuals with impaired glucose tolerance. PLoS ONE 3(8):e2941

Marston OJ, Garfield AS, Heisler LK (2011) Role of central serotonin and melanocortin systems in the control of energy balance. Eur J Pharmacol 660(1):70–79

Marzullo P, Verti B, Savia G, Walker GE, Guzzaloni G, Tagliaferri M, Di Blasio A, Liuzzi A (2004) The relationship between active ghrelin levels and human obesity involves alterations in resting energy expenditure. J Clin Endocrinol Metab 89(2):936–939

Matsumoto M, Hosoda H, Kitajima Y, Morozumi N, Minamitake Y, Tanaka S, Matsuo H, Kojima M, Hayashi Y, Kangawa K (2001a) Structure-activity relationship of ghrelin: pharmacological study of ghrelin peptides. Biochem Biophys Res Commun 287(1):142–146

Matsumoto M, Kitajima Y, Iwanami T, Hayashi Y, Tanaka S, Minamitake Y, Hosoda H, Kojima M, Matsuo H, Kangawa K (2001b) Structural similarity of ghrelin derivatives to peptidyl growth hormone secretagogues. Biochem Biophys Res Commun 284(3):655–659

Melis MR, Mascia MS, Succu S, Torsello A, Muller EE, Deghenghi R, Argiolas A (2002) Ghrelin injected into the paraventricular nucleus of the hypothalamus of male rats induces feeding but not penile erection. Neurosci Lett 329(3):339–343

Mikhail N (2009) The metabolic syndrome: insulin resistance. Curr Hypertens Rep 11(2):156–158

Miwa H, Koseki J, Oshima T, Kondo T, Tomita T, Watari J, Matsumoto T, Hattori T, Kubota K, Iizuka S (2010) Rikkunshito, a traditional Japanese medicine, may relieve abdominal symptoms in rats with experimental esophagitis by improving the barrier function of epithelial cells in esophageal mucosa. J Gastroenterol 45(5):478–487

Montoya-Flores D, Mora O, Tamariz E, Gonzalez-Davalos L, Gonzalez-Gallardo A, Antaramian A, Shimada A, Varela-Echavarria A, Romano-Munoz JL (2012) Ghrelin stimulates myogenic differentiation in a mouse muscle satellite cell line and in primary cultures of bovine myoblasts. J Anim Physiol Anim Nutr (Berl) 96(4):725–738

Moreira FA, Crippa JA (2009) The psychiatric side-effects of rimonabant. Rev Bras Psiquiatr 31(2):145–153

Moulin A, Ryan J, Martinez J, Fehrentz JA (2007) Recent developments in ghrelin receptor ligands. ChemMedChem 2(9):1242–1259

Moulin A, Demange L, Ryan J, M'Kadmi C, Galleyrand JC, Martinez J, Fehrentz JA (2008a) Trisubstituted 1, 2, 4-triazoles as ligands for the ghrelin receptor: on the significance of the orientation and substitution at position 3. Bioorg Med Chem Lett 18(1):164–168

Moulin A, Demange L, Ryan J, Mousseaux D, Sanchez P, Berge G, Gagne D, Perrissoud D, Locatelli V, Torsello A, Galleyrand JC, Fehrentz JA, Martinez J (2008b) New trisubstituted 1, 2, 4-triazole derivatives as potent ghrelin receptor antagonists 3. Synthesis and pharmacological in vitro and in vivo evaluations. J Med Chem 51(3):689–693

Muccioli G, Baragli A, Granata R, Papotti M, Ghigo E (2007) Heterogeneity of ghrelin/growth hormone secretagogue receptors. Toward the understanding of the molecular identity of novel ghrelin/GHS receptors. Neuroendocrinology 86(3):147–164

Nakazato M, Murakami N, Date Y, Kojima M, Matsuo H, Kangawa K, Matsukura S (2001) A role for ghrelin in the central regulation of feeding. Nature 409(6817):194–198

Naleid AM, Grace MK, Cummings DE, Levine AS (2005) Ghrelin induces feeding in the mesolimbic reward pathway between the ventral tegmental area and the nucleus accumbens. Peptides 26(11):2274–2279

Narayanan NS, Guarnieri DJ, DiLeone RJ (2010) Metabolic hormones, dopamine circuits, and feeding. Front Neuroendocrinol 31(1):104–112

Nass R, Gaylinn BD, Thorner MO (2011) The ghrelin axis in disease: potential therapeutic indications. Mol Cell Endocrinol 340(1):106–110

Nathan PJ, O'Neill BV, Napolitano A, Bullmore ET (2011) Neuropsychiatric adverse effects of centrally acting antiobesity drugs. CNS Neurosci Ther 17(5):490–505

Nguyen N, Champion JK, Ponce J, Quebbemann B, Patterson E, Pham B, Raum W, Buchwald JN, Segato G, Favretti F (2012) A review of unmet needs in obesity management. Obes Surg 22(6):956–966

Nishi Y, Hiejima H, Hosoda H, Kaiya H, Mori K, Fukue Y, Yanase T, Nawata H, Kangawa K, Kojima M (2005) Ingested medium-chain fatty acids are directly utilized for the acyl modification of ghrelin. Endocrinology 146(5):2255–2264

Ogden CL, Carroll MD, Kit BK, Flegal KM (2012) Prevalence of obesity and trends in body mass index among US children and adolescents, 1999–2010. JAMA 307(5):483–490

Ogiso K, Asakawa A, Amitani H, Inui A (2011) Ghrelin and anorexia nervosa: a psychosomatic perspective. Nutrition 27(10):988–993

Ohgusu H, Shirouzu K, Nakamura Y, Nakashima Y, Ida T, Sato T, Kojima M (2009) Ghrelin O-acyltransferase (GOAT) has a preference for n-hexanoyl-CoA over n-octanoyl-CoA as an acyl donor. Biochem Biophys Res Commun 386(1):153–158

Olszewski PK, Grace MK, Billington CJ, Levine AS (2003a) Hypothalamic paraventricular injections of ghrelin: effect on feeding and c-Fos immunoreactivity. Peptides 24(6):919–923

Olszewski PK, Li D, Grace MK, Billington CJ, Kotz CM, Levine AS (2003b) Neural basis of orexigenic effects of ghrelin acting within lateral hypothalamus. Peptides 24(4):597–602

Overduin J, Figlewicz DP, Bennett-Jay J, Kittleson S, Cummings DE (2012) Ghrelin increases the motivation to eat, but does not alter food palatability. Am J Physiol Regul Integr Comp Physiol 303(3):R259–R269

Pandit R, de Jong JW, Vanderschuren LJ, Adan RA (2011) Neurobiology of overeating and obesity: the role of melanocortins and beyond. Eur J Pharmacol 660(1):28–42

Panetta R, Greenwood MT (2008) Physiological relevance of GPCR oligomerization and its impact on drug discovery. Drug Discov Today 13(23–24):1059–1066

Parylak SL, Koob GF, Zorrilla EP (2011) The dark side of food addiction. Physiol Behav 104(1):149–156

Patterson M, Murphy KG, le Roux CW, Ghatei MA, Bloom SR (2005) Characterization of ghrelin-like immunoreactivity in human plasma. J Clin Endocrinol Metab 90(4):2205–2211

Patterson M, Bloom SR, Gardiner JV (2011) Ghrelin and appetite control in humans–potential application in the treatment of obesity. Peptides 32(11):2290–2294

Perello M, Zigman JM (2012) The role of ghrelin in reward-based eating. Biol Psychiatry 72(5):347–353

Perello M, Sakata I, Birnbaum S, Chuang JC, Osborne-Lawrence S, Rovinsky SA, Woloszyn J, Yanagisawa M, Lutter M, Zigman JM (2010) Ghrelin increases the rewarding value of high-fat diet in an orexin-dependent manner. Biol Psychiatry 67(9):880–886

Perreault M, Istrate N, Wang L, Nichols AJ, Tozzo E, Stricker-Krongrad A (2004) Resistance to the orexigenic effect of ghrelin in dietary-induced obesity in mice: reversal upon weight loss. Int J Obes Relat Metab Disord 28(7):879–885

Pfluger PT, Kirchner H, Gunnel S, Schrott B, Perez-Tilve D, Fu S, Benoit SC, Horvath T, Joost HG, Wortley KE, Sleeman MW, Tschop MH (2008) Simultaneous deletion of ghrelin and its receptor increases motor activity and energy expenditure. Am J Physiol Gastrointest Liver Physiol 294(3):G610–G618

Powell AG, Apovian CM, Aronne LJ (2011) New drug targets for the treatment of obesity. Clin Pharmacol Ther 90(1):40–51

Rediger A, Tarnow P, Bickenbach A, Schaefer M, Krude H, Grüters A, Biebermann H (2009) Heterodimerization of Hypothalamic G-Protein-Coupled receptors involved in weight regulation. Obes Facts 2(2):80–86

Rediger A, Piechowski CL, Yi CX, Tarnow P, Strotmann R, Gruters A, Krude H, Schoneberg T, Tschop MH, Kleinau G, Biebermann H (2011) Mutually opposite signal modulation by hypothalamic heterodimerization of ghrelin and melanocortin-3 receptors. J Biol Chem 286(45):39623–39631

Richardson NR, Gratton A (1998) Changes in medial prefrontal cortical dopamine levels associated with response-contingent food reward: an electrochemical study in rat. J Neurosci 18(21):9130–9138

Rindi G, Necchi V, Savio A, Torsello A, Zoli M, Locatelli V, Raimondo F, Cocchi D, Solcia E (2002) Characterisation of gastric ghrelin cells in man and other mammals: studies in adult and fetal tissues. Histochem Cell Biol 117(6):511–519

Rodriguez A, Gomez-Ambrosi J, Catalan V, Gil MJ, Becerril S, Sainz N, Silva C, Salvador J, Colina I, Fruhbeck G (2009) Acylated and desacyl ghrelin stimulate lipid accumulation in human visceral adipocytes. Int J Obes (Lond) 33(5):541–552

Rozenfeld R, Devi LA (2010) Receptor heteromerization and drug discovery. Trends Pharmacol Sci 31(3):124–130

Rozenfeld R, Devi LA (2011) Exploring a role for heteromerization in GPCR signalling specificity. Biochem J 433(1):11–18

Rudolph J, Esler WP, O'Connor S, Coish PD, Wickens PL, Brands M, Bierer DE, Bloomquist BT, Bondar G, Chen L, Chuang CY, Claus TH, Fathi Z, Fu W, Khire UR, Kristie JA, Liu XG, Lowe DB, McClure AC, Michels M, Ortiz AA, Ramsden PD, Schoenleber RW, Shelekhin TE, Vakalopoulos A, Tang W, Wang L, Yi L, Gardell SJ, Livingston JN, Sweet LJ, Bullock

WH (2007) Quinazolinone derivatives as orally available ghrelin receptor antagonists for the treatment of diabetes and obesity. J Med Chem 50(21):5202–5216

Sakata I, Nakamura K, Yamazaki M, Matsubara M, Hayashi Y, Kangawa K, Sakai T (2002) Ghrelin-producing cells exist as two types of cells, closed- and opened-type cells, in the rat gastrointestinal tract. Peptides 23(3):531–536

Salim K, Fenton T, Bacha J, Urien-Rodriguez H, Bonnert T, Skynner HA, Watts E, Kerby J, Heald A, Beer M, McAllister G, Guest PC (2002) Oligomerization of G-protein-coupled receptors shown by selective co-immunoprecipitation. J Biol Chem 277(18):15482–15485

Sato T, Fukue Y, Teranishi H, Yoshida Y, Kojima M (2005) Molecular forms of hypothalamic ghrelin and its regulation by fasting and 2-deoxy-d-glucose administration. Endocrinology 146(6):2510–2516

Schellekens H, Dinan TG, Cryan JF (2009) Lean mean fat reducing "ghrelin" machine: hypothalamic ghrelin and ghrelin receptors as therapeutic targets in obesity. Neuropharmacology 58(1):2–16

Schellekens H, Finger BC, Dinan TG, Cryan JF (2012) Ghrelin signalling and obesity: at the interface of stress, mood and food reward. Pharmacol Ther 135(3):316–326

Schellekens H, Dinan TG, Cryan JF (2013a) Ghrelin at the interface of obesity and reward. Vitam Horm 91:285–323

Schellekens H, Dinan TG, Cryan JF (2013b) Taking two to tango: a role for ghrelin receptor heterodimerization in stress and reward. Frontiers Neurosci 7

Schellekens H, van Oeffelen WE, Dinan TG, Cryan JF (2013c) Promiscuous dimerization of the growth hormone secretagogue receptor (GHS-R1a) attenuates ghrelin-mediated signaling. J Biol Chem 288(1):181–191

Seim I, Collet C, Herington AC, Chopin LK (2007) Revised genomic structure of the human ghrelin gene and identification of novel exons, alternative splice variants and natural antisense transcripts. BMC Genom 8:298

Shiiya T, Nakazato M, Mizuta M, Date Y, Mondal MS, Tanaka M, Nozoe S, Hosoda H, Kangawa K, Matsukura S (2002) Plasma ghrelin levels in lean and obese humans and the effect of glucose on ghrelin secretion. J Clin Endocrinol Metab 87(1):240–244

Shimbara T, Mondal MS, Kawagoe T, Toshinai K, Koda S, Yamaguchi H, Date Y, Nakazato M (2004) Central administration of ghrelin preferentially enhances fat ingestion. Neurosci Lett 369(1):75–79

Shrestha YB, Wickwire K, Giraudo S (2009) Effect of reducing hypothalamic ghrelin receptor gene expression on energy balance. Peptides 30(7):1336–1341

Shuto Y, Shibasaki T, Otagiri A, Kuriyama H, Ohata H, Tamura H, Kamegai J, Sugihara H, Oikawa S, Wakabayashi I (2002) Hypothalamic growth hormone secretagogue receptor regulates growth hormone secretion, feeding, and adiposity. J Clin Invest 109(11):1429–1436

Simpson K, Martin A, Niamh M, Bloom SR (2008) Hypothalamic regulation of appetite. Expert Rev Endocrinol Metab 3(5):577–592

Simpson KA, Martin NM, Bloom SR (2009) Hypothalamic regulation of food intake and clinical therapeutic applications. Arq Bras Endocrinol Metabol 53(2):120–128

Skibicka KP, Dickson SL (2011) Ghrelin and food reward: the story of potential underlying substrates. Peptides 32(11):2265–2273

Skibicka KP, Hansson C, Egecioglu E, Dickson SL (2012) Role of ghrelin in food reward: impact of ghrelin on sucrose self-administration and mesolimbic dopamine and acetylcholine receptor gene expression. Addict Biol 17(1):95–107

Smith NJ, Milligan G (2010) Allostery at G protein-coupled receptor homo-and heteromers: uncharted pharmacological landscapes. Pharmacol Rev 62(4):701–725

Soares JB, Leite-Moreira AF (2008) Ghrelin, des-acyl ghrelin and obestatin: three pieces of the same puzzle. Peptides 29(7):1255–1270

Soares JB, Roncon-Albuquerque R Jr, Leite-Moreira A (2008) Ghrelin and ghrelin receptor inhibitors: agents in the treatment of obesity. Expert Opin Ther Targets 12(9):1177–1189

Soriano-Guillen L, Barrios V, Campos-Barros A, Argente J (2004) Ghrelin levels in obesity and anorexia nervosa: effect of weight reduction or recuperation. J Pediatr 144(1):36–42

Stanley S, Wynne K, McGowan B, Bloom S (2005) Hormonal regulation of food intake. Physiol Rev 85(4):1131–1158

Stengel A, Goebel M, Wang L, Tache Y (2010) Ghrelin, des-acyl ghrelin and nesfatin-1 in gastric X/A-like cells: role as regulators of food intake and body weight. Peptides 31(2):357–369

Stricker-Krongrad A, Richy S, Beck B (2002) Orexins/hypocretins in the ob/ob mouse: hypothalamic gene expression, peptide content and metabolic effects. Regul Pept 104(1–3):11–20

Sun Y, Ahmed S, Smith RG (2003) Deletion of ghrelin impairs neither growth nor appetite. Mol Cell Biol 23(22):7973–7981

Sun Y, Wang P, Zheng H, Smith RG (2004) Ghrelin stimulation of growth hormone release and appetite is mediated through the growth hormone secretagogue receptor. Proc Natl Acad Sci USA 101(13):4679–4684

Sun Y, Butte NF, Garcia JM, Smith RG (2008) Characterization of adult ghrelin and ghrelin receptor knockout mice under positive and negative energy balance. Endocrinology 149(2):843–850

Suzuki K, Simpson KA, Minnion JS, Shillito JC, Bloom SR (2010) The role of gut hormones and the hypothalamus in appetite regulation. Endocr J 57(5):359–372

Swinburn BA, Sacks G, Hall KD, McPherson K, Finegood DT, Moodie ML, Gortmaker SL (2011) The global obesity pandemic: shaped by global drivers and local environments. The Lancet 378(9793):804–814

Tamura H, Kamegai J, Shimizu T, Ishii S, Sugihara H, Oikawa S (2002) Ghrelin stimulates GH but not food intake in arcuate nucleus ablated rats. Endocrinology 143(9):3268–3275

Tanaka M, Hayashida Y, Nakao N, Nakai N, Nakashima K (2001) Testis-specific and developmentally induced expression of a ghrelin gene-derived transcript that encodes a novel polypeptide in the mouse. Biochim Biophys Acta 1522(1):62–65

Teitler M, Klein MT (2012) A new approach for studying GPCR dimers: drug-induced inactivation and reactivation to reveal GPCR dimer function in vitro, in primary culture, and in vivo. Pharmacol Ther 133(2):205–217

Terrillon S, Bouvier M (2004) Roles of G-protein-coupled receptor dimerization. EMBO Rep 5(1):30–34

Tisdale MJ (1997) Biology of cachexia. J Natl Cancer Inst 89(23):1763–1773

Tolle V, Low MJ (2008) In vivo evidence for inverse agonism of Agouti-related peptide in the central nervous system of proopiomelanocortin-deficient mice. Diabetes 57(1):86–94

Toshinai K, Date Y, Murakami N, Shimada M, Mondal MS, Shimbara T, Guan JL, Wang QP, Funahashi H, Sakurai T, Shioda S, Matsukura S, Kangawa K, Nakazato M (2003) Ghrelin-induced food intake is mediated via the orexin pathway. Endocrinology 144(4):1506–1512

Tovar S, Nogueiras R, Tung LY, Castaneda TR, Vazquez MJ, Morris A, Williams LM, Dickson SL, Dieguez C (2005) Central administration of resistin promotes short-term satiety in rats. Eur J Endocrinol 153(3):R1–R5

Tschop M, Wawarta R, Riepl RL, Friedrich S, Bidlingmaier M, Landgraf R, Folwaczny C (2001a) Post-prandial decrease of circulating human ghrelin levels. J Endocrinol Invest 24(6):RC19–RC21

Tschop M, Smiley DL, Heiman ML (2000) Ghrelin induces adiposity in rodents. Nature 407(6806):908–913

Tschop M, Weyer C, Tataranni PA, Devanarayan V, Ravussin E, Heiman ML (2001b) Circulating ghrelin levels are decreased in human obesity. Diabetes 50(4):707–709

Ukkola O, Ravussin E, Jacobson P, Snyder EE, Chagnon M, Sjostrom L, Bouchard C (2001) Mutations in the preproghrelin/ghrelin gene associated with obesity in humans. J Clin Endocrinol Metab 86(8):3996–3999

Venkova K, Greenwood-Van Meerveld B (2008) Application of ghrelin to gastrointestinal diseases. Curr Opin Investig Drugs 9(10):1103–1107

Vivenza D, Rapa A, Castellino N, Bellone S, Petri A, Vacca G, Aimaretti G, Broglio F, Bona G (2004) Ghrelin gene polymorphisms and ghrelin, insulin, IGF-I, leptin and anthropometric data in children and adolescents. Eur J Endocrinol 151(1):127–133

Volkow ND, Wang GJ, Baler RD (2010) Reward, dopamine and the control of food intake: implications for obesity. Trends Cogn Sci 15(1):37–46

Volkow ND, Wang GJ, Fowler JS, Tomasi D, Baler R, Carter CS, Dalley JW (2012) Food and drug reward: overlapping circuits in human obesity and addiction. Brain imaging in behavioral neuroscience; current topics in behavioral neuroscience, vol 11. Springer, Berlin, pp 1–24

Wise RA (2006) Role of brain dopamine in food reward and reinforcement. Philos Trans R Soc Lond B Biol Sci 361(1471):1149–1158

Wise RA, Rompre PP (1989) Brain dopamine and reward. Annu Rev Psychol 40:191–225

World_Health_Organisation (2013) http://www.who.int/topics/obesity/en/index.html

Wortley KE, Anderson KD, Garcia K, Murray JD, Malinova L, Liu R, Moncrieffe M, Thabet K, Cox HJ, Yancopoulos GD, Wiegand SJ, Sleeman MW (2004) Genetic deletion of ghrelin does not decrease food intake but influences metabolic fuel preference. Proc Natl Acad Sci USA 101(21):8227–8232

Wortley KE, del Rincon JP, Murray JD, Garcia K, Iida K, Thorner MO, Sleeman MW (2005) Absence of ghrelin protects against early-onset obesity. J Clin Invest 115(12):3573–3578

Wren AM, Seal LJ, Cohen MA, Brynes AE, Frost GS, Murphy KG, Dhillo WS, Ghatei MA, Bloom SR (2001) Ghrelin enhances appetite and increases food intake in humans. J Clin Endocrinol Metab 86(12):5992–5995

Xu Y, Elmquist JK, Fukuda M (2011) Central nervous control of energy and glucose balance: focus on the central melanocortin system. Ann N Y Acad Sci 1243:1–14

Yamanaka A, Beuckmann CT, Willie JT, Hara J, Tsujino N, Mieda M, Tominaga M, Yagami K, Sugiyama F, Goto K, Yanagisawa M, Sakurai T (2003) Hypothalamic orexin neurons regulate arousal according to energy balance in mice. Neuron 38(5):701–713

Yang J, Brown MS, Liang G, Grishin NV, Goldstein JL (2008a) Identification of the acyltransferase that octanoylates ghrelin, an appetite-stimulating peptide hormone. Cell 132(3):387–396

Yang J, Zhao TJ, Goldstein JL, Brown MS (2008b) Inhibition of ghrelin O-acyltransferase (GOAT) by octanoylated pentapeptides. Proc Natl Acad Sci U S A 105(31):10750–10755

Yi CX, Heppner K, Tschop MH (2011) Ghrelin in eating disorders. Mol Cell Endocrinol 340(1):29–34

Yildiz BO, Suchard MA, Wong ML, McCann SM, Licinio J (2004) Alterations in the dynamics of circulating ghrelin, adiponectin, and leptin in human obesity. Proc Natl Acad Sci USA 101(28):10434–10439

Yu M, Lizarzaburu M, Beckmann H, Connors R, Dai K, Haller K, Li C, Liang L, Lindstrom M, Ma J, Motani A, Wanska M, Zhang A, Li L, Medina JC (2010) Identification of piperazine-bisamide GHSR antagonists for the treatment of obesity. Bioorg Med Chem Lett 20(5):1758–1762

Zhang Y, Proenca R, Maffei M, Barone M, Leopold L, Friedman JM (1994) Positional cloning of the mouse obese gene and its human homologue. Nature 372(6505):425–432

Zhi J, Melia AT, Eggers H, Joly R, Patel IH (1995) Review of limited systemic absorption of orlistat, a lipase inhibitor, in healthy human volunteers. J Clin Pharmacol 35(11):1103–1108

Zigman JM, Nakano Y, Coppari R, Balthasar N, Marcus JN, Lee CE, Jones JE, Deysher AE, Waxman AR, White RD, Williams TD, Lachey JL, Seeley RJ, Lowell BB, Elmquist JK (2005) Mice lacking ghrelin receptors resist the development of diet-induced obesity. J Clin Invest 115(12):3564–3572

Zigman JM, Jones JE, Lee CE, Saper CB, Elmquist JK (2006) Expression of ghrelin receptor mRNA in the rat and the mouse brain. J Comp Neurol 494(3):528–548

Zorrilla EP, Iwasaki S, Moss JA, Chang J, Otsuji J, Inoue K, Meijler MM, Janda KD (2006) Vaccination against weight gain. Proc Natl Acad Sci USA 103(35):13226–13231

Ghrelin Receptor Antagonism as a Potential Therapeutic Target for Alcohol Use Disorders: A Preclinical Perspective

Elisabet Jerlhag and Jörgen A. Engel

Abstract The rewarding properties of natural and chemical reinforcers are mediated via the reward systems, such as the cholinergic-dopaminergic reward link. A dysfunction in these reward systems underlies development of addictive behaviours such as alcohol use disorder. By elucidating the complex neurobiological mechanisms involved in the drug-induced activation of the mesolimbic dopamine system, novel treatment strategies can be identified. Recent work has suggested that the gut–brain peptide ghrelin may be such candidates. Indeed, the orexigenic peptide ghrelin activates the cholinergic-dopaminergic reward link. Ghrelin may thereby increase the incentive salience for motivated behaviours such as reward seeking. Moreover, preclinical findings show that ghrelin signalling is required for reward induced by alcohol, for the motivation to consume alcohol and for the intake of alcohol in rodents. Reward induced by other additive drugs such as nicotine, cocaine and amphetamine also involve ghrelin and its receptor. Human genetic data support a role for ghrelin in drug reward. Polymorphisms in ghrelin-related genes are associated with increased alcohol intake, smoking as well as amphetamine dependence in humans. Furthermore, plasma levels of ghrelin are associated with alcohol dependence as well as with craving. Finally, another gut–brain peptide known to regulate food intake, i.e., the anorectic peptide glucagone-like-peptide-1 (GLP-1), was recently shown to regulate drug reinforcement. Peripheral treatment with a GLP-1 analogue attenuated alcohol-induced reward as well as decreased alcohol intake and alcohol seeking behaviour in rodents. In addition, GLP-1 analogues appear to attenuate drug-induced reward. Collectively, these data suggest that ghrelin and GLP-1 receptors may be novel targets for development of pharmacological treatments of addictive behaviours such as alcohol dependence.

Keywords Ghrelin, GLP-1 · Reward · Addiction · Dopamine · Alcohol · Nicotine · Gut–brain peptides

E. Jerlhag (✉) · J. A. Engel
Department of Pharmacology, Institute of Neuroscience and Physiology, The Sahlgrenska Academy at the University of Gothenburg, PO Box 431, SE-405 30 Gothenburg, Sweden
e-mail: Elisabet.jerlhag@pharm.gu.se

J. Portelli and I. Smolders (eds.), *Central Functions of the Ghrelin Receptor,*
The Receptors 25, DOI: 10.1007/978-1-4939-0823-3_7,
© Springer Science+Business Media New York 2014

Addictive Behaviours and the Cholinergic-Dopaminergic Reward Link

The rewarding properties of natural and chemical reinforcers are mediated via the reward systems in the brain (Damsma et al. 1992; Engel et al. 1988; Wise and Rompre 1989). These evolutionary conserved systems also appear to enhance the motivation for behaviours that increase the probability of survival such as food seeking (Berridge and Robinson 1998; Robinson and Berridge 1993). An important part of these reward systems is the cholinergic-dopaminergic reward link, which encompasses the cholinergic afferent projection from the laterodorsal tegmental area (LDTg) onto the ventral tegmental area (VTA) dopamine cells together with the mesolimbic dopamine system [i.e., the dopamine neurons from the VTA to nucleus accumbens (NAc)] (Larsson and Engel 2004). Dependence producing drugs and natural rewards, in addition to increasing dopamine release in the NAc, simultaneously enhance the acetylcholine levels in the VTA (Lanca et al. 2000; Larsson et al. 2005; Rada et al. 2000; Yeomans et al. 1993) suggesting that this link has an important role in the reward regulation.

Human imaging studies have revealed that there is an underlying disruption in the reward systems in individuals with drug addiction such as alcohol use disorder (AUD) and smoking (Holden 2001; Potenza et al. 2003; Volkow et al. 2003a, b). Smoking, AUD and other chemical addictions are chronic, relapsing brain disorders and they cause a wide range of serious effects to the individual as well as the society. Addiction is therefore considered to be one of our societies major public health problems (Koob and Le Moal 2001; Duaso and Duncan 2012). Recent studies have shown that there are behavioural parallels, e.g. loss of control, between chemical addiction and 'behavioural' addictions, such as compulsive overeating, compulsive shopping and gambling (Davis and Woodside 2002). Most interestingly, individuals with 'behavioural' addictions such as compulsive over-eating display similar disruption in the reward systems as patients with drug dependence (Volkow et al. 2003b). 'Behavioural' addictions have therefore been included in the definition of addiction and are together with drug dependence called addictive behaviours.

Development of addiction depends, at least in part, on the effects of drugs of abuse on the mesolimbic dopamine system [for review see (Larsson and Engel 2004; Soderpalm et al. 2009; Tupala and Tiihonen 2004; Volkow and Li 2004)], although several neurotransmitter systems collectively orchestrate the reward profile of drugs of abuse. The clinical efficacy of the available pharmaceutical agents for addictive behaviours such as AUD is limited (Anton et al. 2006) and there is, therefore, a need for novel treatment strategies. By elucidating the complex neurobiological mechanisms involved in the drug-induced activation of the mesolimbic dopamine system, novel treatment strategies can be identified.

A variety of human studies suggest that common neurobiological mechanisms underlie different forms of addictive behaviours, including AUD, smoking, other forms of chemical addiction as well as compulsive overeating (Thiele et al. 2003;

Morganstern et al. 2011). This raises the possibility that endocrine signals from the gut traditionally known to regulate food intake, energy and body weight homeostasis, such as ghrelin, may play an important role in reward regulation as well as in development of drug dependence. The role of ghrelin signalling in reward as well as in drug-mediated behaviours is reviewed herein (vide infra).

Ghrelin Activates the Cholinergic-Dopaminergic Reward Link

While previous research indicates that circulating ghrelin has physiological roles for food intake, appetite as well as meal initiation (Wren et al. 2000, 2001a, b; Egecioglu et al. 2011), the findings that growth hormone secretagogue receptors (ghrelin receptors) are expressed throughout the mesolimbic dopamine system (Guan et al. 1997; Zigman et al. 2006) raised the hypothesis that ghrelin may have a role in reward regulation. Initially, it was shown that central administration of ghrelin induces an increase in accumbal dopamine release and also induces locomotor stimulation in mice (Jerlhag et al. 2006), indicating that ghrelin activates the mesolimbic dopamine system. In support of this are the findings demonstrating that ghrelin administration into the VTA or LDTg (important reward nodes) induces an increase in accumbal dopamine release as well as a locomotor stimulation (Jerlhag et al. 2006, 2007) and that local VTA administration increases dopamine turnover in N.Acc (Abizaid et al. 2006). Recent data showed that ghrelin administered locally into the LDTg or peripherally concomitantly increases ventral tegmental acetylcholine as well as accumbal dopamine release and this synchronous neurotransmitter release is blocked by a ghrelin receptor 1a antagonist (Jerlhag et al. 2012). Collectively, these data suggest that ghrelin activates the cholinergic-dopaminergic reward link via direct actions in the LDTg as well as the VTA. In support of this are the findings showing that ghrelin receptor 1a is expressed on a sub-population of dopamine cells in the VTA (Abizaid et al. 2006) as well as on cholinergic neurons in the LDTg (Dickson et al. 2010). In addition, it should be noticed that NAc may be involved in ghrelin-mediated reward since ghrelin receptor 1a is expressed in this reward area (Landgren et al. 2011a).

Previously, it was shown that $\alpha3\beta2$, $\beta3$ and $\alpha6$ nicotinic acetylcholine receptors subtypes in the VTA appear to be critical for the ability of ghrelin to activate the cholinergic-dopaminergic reward link (Jerlhag et al. 2008). Neurochemical analogies between ghrelin and alcohol could therefore be implied since $\alpha3\beta2$, $\beta3$ and $\alpha6$ nicotinic acetylcholine receptors subtypes in the VTA mediate the reinforcing properties of alcohol (Larsson et al. 2005, 2004; Lof et al. 2007; Steensland et al. 2007; Salome et al. 2009). These data are verified in clinical tests; thus blocking these subtypes reduces the intake of alcohol in heavy drinking smokers in a laboratory setting as well as in a double-blinded clinical trial (McKee et al. 2009; Mitchell et al. 2012) and one haplotype of the $\alpha6$ gene is associated with heavy alcohol use (Landgren et al. 2009). In addition, local perfusion of the unselective nicotinic antagonist mecamylamine into the VTA blocks the ability of ghrelin (into the LDTg) to increase NAc-dopamine, but not the increase of VTA-acetylcholine (Jerlhag et al. 2012). Taken together, this provides proof of concept that ghrelin activates ghrelin receptor 1a in LDTg causing a release of acetylcholine in the VTA, which activates local nicotinic acetylcholine receptors (specifically $\alpha3\beta2$, $\beta3$ and $\alpha6$ subtypes) causing a release of accumbal dopamine. Given that the cholinergic-dopaminergic reward link mediates the incentive salience of motivated behaviours, the present data collectively imply that ghrelin, via activation of this reward link, mediates motivated behaviours such as reward seeking.

In addition to the cholinergic afferent to the VTA, the activity of dopaminergic neurons in the VTA are regulated via various other afferents. Indeed, ghrelin receptor 1a within the VTA is present, not only on the dopaminergic cells, but also on pre-synaptic afferents such as, e.g. GABAergic interneurons (Abizaid et al. 2006). The possibility that these could mediate the ability of ghrelin to activate the reward systems should therefore be considered. This is supported by the findings that a non-selective glutamate NMDA receptor antagonist (AP5), but not an opioid receptor antagonist or an orexin receptor A antagonist, blocks the ability of ghrelin to activate the reward systems as measured by increasing the locomotor activity, accumbal dopamine release and condition a place preference (Jerlhag et al. 2011a). Given that hyperghrelinemia in association with addictive behaviours (see "Ghrelin and Sleep Regulation" by Leggio and Feduccia), future therapeutic targets for these disorders may include agents such as nicotinic acetylcholine receptor or glutamate receptor antagonists acting at the level of the cholinergic-dopaminergic reward link.

The findings that ghrelin is produced in the gastrointestinal tract (Kojima et al. 1999) and that this gut–brain hormone passes the blood–brain barrier (Banks et al. 2002) raise the possibility that circulating endogenous ghrelin may reach brain reward nodes and thereby induce reward. In support of this hypothesis are the findings showing that peripherally administered ghrelin increases accumbal dopamine release (Jerlhag 2008), specifically in the shell region of NAc (Quarta et al. 2009) as well as induces a locomotor stimulation in addition to a conditioned place preference (Jerlhag 2008). Furthermore, local administration of ghrelin receptor 1a antagonists in the VTA blocked peripherally administered ghrelin to increase food intake and to induce reward in rodents (Abizaid et al. 2006; Jerlhag et al. 2011a), showing that ghrelin targets the dopamine system directly via ghrelin receptor 1a in the VTA. Moreover, imaging data revealed that peripheral ghrelin administration causes a focal activation of a network of VTA, NAc and lateral hypothalamus in rats (Wellman et al. 2012). This is further substantiated by human functional magnetic resonance imaging data showing that ghrelin administration to healthy volunteers alters the brain response to visual food cues in reward-related areas such as the NAc (Malik et al. 2008). Given that accumbal dopamine release appears to mediate the rewarding properties of incentives (Robinson and Berridge 1993; Wise and Bozarth 1987; Engel et al. 1988), the collective data suggest that ghrelin have a direct role in reward regulation.

Ghrelin Signalling is Required for Alcohol-Mediated Behaviours: A Preclinical Perspective

The contention that common neurobiological mechanisms underlie different forms of addictive behaviours (Thiele et al. 2003; Morganstern et al. 2011) is further supported by our findings that ghrelin signalling is required for alcohol-induced reward, alcohol intake and for the motivation to consume alcohol (vide infra).

In support of this notion are the findings showing that suppressed ghrelin signalling, with either pharmacological (central or peripheral administration) or genetical approaches, reduces the rewarding properties of alcohol as measured by locomotor stimulation, accumbal dopamine release and conditioned place preference (Jerlhag et al. 2009, 2011b). In addition, peripheral or central administration of the ghrelin receptor 1a antagonists (JMV2959 or BIM28163 respectively) reduces the intake of alcohol in mice for 12 weeks (Jerlhag et al. 2009). In accordance are the findings that the ghrelin receptor 1a antagonist JMV2959 reduces high-alcohol consumption in high-alcohol consuming Wistar as well as in alcohol-preferring (AA) rats (Landgren et al. 2012). A recent study supports these data since it was shown that ghrelin receptor 1a treatment decreases alcohol intake in rats exposed to alcohol for 2, 5 and 10 months (Suchankova et al. 2013). Specifically, it was shown that the ability of acute ghrelin receptor 1a treatment to reduce alcohol intake was more pronounced after 5, compared to 2 months of alcohol exposure. In addition, repeated JMV2959 treatment decreased alcohol intake without inducing tolerance or rebound increase in alcohol intake after the treatment. In addition, the ghrelin receptor 1a antagonist prevented the alcohol deprivation effect, an important characteristic of alcohol dependence, in rats. In the rat exposed to alcohol for 10 months there was a significant down-regulation of the ghrelin receptor 1a expression in the VTA in high-compared to low-alcohol consuming rats. No differences in methylation degree were found in high-compared to low-alcohol consuming rats (Suchankova et al. 2013). The independent findings showing that another ghrelin receptor 1a antagonist (D-Lys3-GHRP-6) reduces alcohol intake in rats (Kaur and Ryabinin 2010), supports that ghrelin signalling can regulate alcohol intake. A role of ghrelin in alcohol consumption regulation is supported by human genetic and clinical findings (see "Clinical Research on the Ghrelin Axis and Alcohol Consumption" by Leggio and Feduccia). Another important part of AUD is the motivational properties of alcohol intake and it was recently shown that peripheral administration of a ghrelin receptor 1a antagonist reduces the motivation to consume alcohol as measured by operant self-administration in rats (Landgren et al. 2012). Previous studies have proposed that the ability of ghrelin to increase food intake are mediated via hypothalamic ghrelin receptor 1a (Wren et al. 2000). However, ghrelin receptor 1a in this area does not appear to be important for alcohol-mediated behaviours since hypothalamic administration of ghrelin does not influence the intake of alcohol in rats (Schneider et al. 2007). However, ghrelin adbministration into reward areas known to express ghrelin receptor 1a, i.e. the LDTg or VTA, increases the intake of alcohol in mice (Jerlhag et al. 2009), implying that ghrelin signalling within the mesolimbic dopamine system is important for alcohol-mediated behaviours. Furthermore, the findings that peripheral ghrelin administration to mice exposed to alcohol for 3 days only slightly increases alcohol intake (Lyons et al. 2008) suggest that ghrelin signalling is more important in rodents exposed to alcohol for longer, rather than shorter, periods of time.

Growing evidence shows that ghrelin increases whereas ghrelin receptor 1a antagonists reduce food intake and appetite in humans as well as in rodents [for review see (Egecioglu et al. 2011)]. The possibility that the effects of ghrelin and ghrelin receptor 1a treatment on alcohol intake could be driven by the caloric value of alcohol rather than by effects on the rewarding properties of alcohol should therefore be considered. However, this appears less likely since animal studies show that the rewarding properties of rewards without caloric content, such as cocaine, amphetamine and nicotine, are attenuated by ghrelin receptor 1a antagonist treatment (vide infra). In addition ghrelin receptor 1a antagonist treatment reduces the intake of saccharine, another reward without calories (Landgren et al. 2011b). Furthermore, ghrelin receptor 1a antagonism suppresses parameters associated with alcohol's rewarding properties rather than its caloric content (Jerlhag et al. 2009).

Ghrelin Signalling is Required for Drug-Induced Reward: A Preclinical Perspective

Given that central ghrelin signalling is required for alcohol-mediated behaviours, the question arose regarding the extent to which this gut–brain hormone could be important for reward regulation, in general, such as reinforcement from other drugs of abuse.

Indeed, peripheral ghrelin administration augments cocaine-induced locomotor stimulation (Wellman et al. 2005) as well as conditioned place preference in rats (Davis et al. 2007). These data are supported by the findings that elevated plasma levels of ghrelin are associated with cocaine-seeking behaviour in rats (Tessari et al. 2007). Moreover, peripheral administration of a ghrelin receptor 1a antagonist attenuates the amphetamine- and cocaine-induced locomotor stimulation, accumbal dopamine release and conditioned place preference (Jerlhag et al. 2010) and genetic or pharmacologic ghrelin receptor 1a antagonism attenuates the cocaine-induced locomotor stimulation as well as sensitization in rats (Clifford et al. 2012; Abizaid et al. 2011). In addition to alcohol and psychostimulant drugs, ghrelin signalling appears to mediate nicotine-induced reward. Thus, ghrelin receptor 1a antagonist blocks the rewarding properties of nicotine as measured by locomotor stimulation, accumbal dopamine and conditioned place preference (Jerlhag and Engel 2011) as well as attenuates development of nicotine-induced locomotor sensitization in rodents (Wellman et al. 2011). In support of a general role for ghrelin in drug-induced reward are the data showing that food restriction, that increases ghrelin levels (Gualillo et al. 2002), augments amphetamine- as well as cocaine-induced locomotor stimulation, increases the self-administration of cocaine or amphetamine and enhances cocaine-seeking behaviour in rats (Carroll et al. 1979). Taken together with human genetic data showing associations between polymorphisms in ghrelin signalling genes and the intake of amphetamine or nicotine (see 'Clinical Research on the Ghrelin Axis and Alcohol Consumption' by Leggio and Feduccia), a role of ghrelin and its receptor in drug-induced reinforcement may be implied.

Other Gut–Brain Peptides Mediate Drug-Induced Reward in Rodents

The notion that common signalling systems regulate the intake of food and alcohol (Thiele et al. 2004) imply that other endocrine signals from the gut than ghrelin may have a role in reward regulation. Indeed, the anorectic peptide, glucagone-like-peptide-1 (GLP-1) was recently shown to regulate drug-induced reward in rodents. Indeed, it was shown that peripheral treatment with the GLP-1 analogue, exendin-4, attenuated alcohol-induced locomotor activity, accumbal dopamine release and conditioned place preference as well as decreased alcohol intake and alcohol-seeking behaviour in rodents (Egecioglu et al. 2012). Furthermore, exendin-4 attenuates amphetamine-induced locomotor stimulation and cocaine-induced conditioned place preference in rodents (Erreger et al. 2012; Graham et al. 2013). In support are the recent data showing that exendin-4, at a dose that has no effet *per se*, attenuates the ability of cocaine as well as amphetamine to induce a locomotor stimulation, accumbal dopamine release and to condition a place preference in mice (Egecioglu et al. 2013). Moreover, gastric bypass, which reduces ghrelin and increases GLP-1 plasma levels, reduces the alcohol intake in both humans and rats (Davis et al. 2012). Another gut–brain peptide mediating drug reward appears to be the hunger hormone galanin, which increases alcohol consumption and is required for nicotine-induced reward in rodents (Lewis et al. 2004; Rada et al. 2004; Neugebauer et al. 2011). Furthermore, animal studies show that the anorectic peptide cholecystokinin reduces alcohol consumption and that a cholecystokinin antagonist reduces nicotine withdrawal (Rasmussen et al. 1996; Kulkosky 1984). Leptin, an adipose-derived hormone reducing food intake, has been shown to reduce alcohol consumption as well as block reward induced by psychostimulant drugs in rodents (Blednov et al. 2004; Opland et al. 2010). Further support for a role of leptin in drug reinforcement are the data showing that elevated plasma level of the leptin is associated with nicotine craving in humans (von der Goltz et al. 2010). Moreover, the plasma level of the hunger peptide orexin is associated with nicotine craving in humans (von der Goltz et al. 2010) and an orexin antagonist blocks reward induced by psychostimulant drugs (Borgland et al. 2006). The findings presented in this chapter, i.e. that the gut–brain hormones are important players of the reward scene, implicate that these peptides have a broader role than just regulating energy homeostasis comprising enhancement of the incentive value of signals of importance for survival such as food seeking. Consequently, interfering with these systems may constitute new targets for development of novel treatment strategies for addictive behaviours such as alcohol use disorders.

Acknowledgment The book chapter was supported by grants from the Swedish Research Council (grant no. K2006-21X-04247-33-3, 2009-2782 and K2010-80X-21496-01-6), The Swedish brain foundation, LUA/ALF (grant no. 148251) from the Sahlgrenska University Hospital, Alcohol research council of the Swedish alcohol retailing monopoly and the foundations of Adlerbertska, Fredrik and Ingrid Thuring, Tore Nilsson, Längmanska, Torsten and Ragnar Söderberg, Wilhelm and Martina Lundgren, NovoNordisk, Knut and Alice Wallenberg, Magnus Bergvall, Anérs, Jeansons, Åke Wiberg, the Swedish Society of Medicine, Swedish Society for Medical Research.

References

Abizaid A, Liu ZW, Andrews ZB, Shanabrough M, Borok E, Elsworth JD, Roth RH, Sleeman MW, Picciotto MR, Tschop MH, Gao XB, Horvath TL (2006) Ghrelin modulates the activity and synaptic input organization of midbrain dopamine neurons while promoting appetite. J Clin Invest 116:3229–3239

Abizaid A, Mineur YS, Roth RH, Elsworth JD, Sleeman MW, Picciotto MR, Horvath TL (2011) Reduced locomotor responses to cocaine in ghrelin-deficient mice. Neuroscience 192:500–506

Anton RF, O'Malley SS, Ciraulo DA, Cisler RA, Couper D, Donovan DM, Gastfriend DR, Hosking JD, Johnson BA, Locastro JS, Longabaugh R, Mason BJ, Mattson ME, Miller WR, Pettinati HM, Randall CL, Swift R, Weiss RD, Williams LD, Zweben A (2006) Combined pharmacotherapies and behavioral interventions for alcohol dependence: the COMBINE study: a randomized controlled trial. JAMA 295:2003–2017

Banks WA, Tschop M, Robinson SM, Heiman ML (2002) Extent and direction of ghrelin transport across the blood-brain barrier is determined by its unique primary structure. J Pharmacol Exp Ther 302:822–827

Berridge KC, Robinson TE (1998) What is the role of dopamine in reward: hedonic impact, reward learning, or incentive salience? Brain Res Rev 28:309–369

Blednov YA, Walker D, Harris RA (2004) Blockade of the leptin-sensitive pathway markedly reduces alcohol consumption in mice. Alcohol Clin Exp Res 28:1683–1692

Borgland SL, Taha SA, Sarti F, Fields HL, Bonci A (2006) Orexin a in the VTA is critical for the induction of synaptic plasticity and behavioral sensitization to cocaine. Neuron 49:589–601

Carroll ME, France CP, Meisch RA (1979) Food-Deprivation Increases Oral and Intravenous Drug Intake in Rats. Science 205:319–321

Clifford pS, Rodriguez J, Schul D, Hughes S, Kniffin T, Hart N, Eitan S, Brunel L, Fehrentz JA, Aartinez J, Wellman PJ (2012) Attenuation of cocaine-induced locomotor sensitization in rats sustaining genetic or pharmacologic antagonism of ghrelin receptors. Addict Biol 17:956–963

Damsma G, Pfaus JG, Wenkstern D, Phillips AG, Fibiger HC (1992) Sexual behavior increases dopamine transmission in the nucleus accumbens and striatum of male rats: comparison with novelty and locomotion. Behav Neurosci 106:181–191

Davis C, Woodside DB (2002) Sensitivity to the rewarding effects of food and exercise in the eating disorders. Compr Psychiatry 43:189–194

Davis JF, Schurdak JD, Magrisso IJ, Mul JD, Grayson BE, Pfluger PT, Tschoep MH, Seeley RJ, Benoit SC (2012) Gastric bypass surgery attenuates ethanol consumption in ethanol-preferring rats. Biol Psychiatry 72:354–360

Davis KW, Wellman PJ, Clifford PS (2007) Augmented cocaine conditioned place preference in rats pretreated with systemic ghrelin. Regul Pept 140:148–152

Dickson SL, Hrabovszky E, Hansson C, Jerlhag E, Alvarez-Crespo M, Skibicka KP, Molnar CS, Liposits Z, Engel JA, Egecioglu E (2010) Blockade of central nicotine acetylcholine receptor signaling attenuate ghrelin-induced food intake in rodents. Neuroscience 171:1180–1186

Duaso M, Duncan D (2012) Health impact of smoking and smoking cessation strategies: current evidence. Br J Community Nurs 17:356–363

Egecioglu E, Skibicka KP, Hansson C, Alvarez-Crespo M, Friberg PA, Jerlhag E, Engel JA, Dickson SL (2011) Hedonic and incentive signals for body weight control. Rev Endocr Metab Disord 12:141–151

Egecioglu E, Steensland P, Fredriksson I, Feltmann K, Engel JA, Jerlhag E (2012) The glucagon-like peptide 1 analogue exendin-4 attenuates alcohol mediated behaviors in rodents. Psychoneuroendocrinology. doi:10.1016/j.psyneuen.2012.11.009

Egecioglu E, Engel JA, Jerlhag E (2013) The glucagon-like peptide 1 analogue exendin-4 attenuates the rewarding properties of psychostimulant drugs in mice. Plos One. doi:10.1371/journal.pone.0069010

Engel JA, Fahlke C, Hulthe P, Hard E, Johannessen K, Snape B, Svensson L (1988) Biochemical and behavioral evidence for an interaction between ethanol and calcium-channel antagonists. Alcohol Alcohol 23:A13–A13

Erreger K, Davis AR, Poe AM, Greig NH, Stanwood GD, Galli A (2012) Exendin-4 decreases amphetamine-induced locomotor activity. Physiol Behav 106:574–578

Graham dL, Erreger K, Galli A, Stanwood GD (2013) GLP-1 analog attenuates cocaine reward. Mol Psychiatry 18:961–962

Gualillo O, Caminos JE, Nogueiras R, Seoane LM, Arvat E, Ghigo E, Casanueva FF, Dieguez C (2002) Effect of food restriction on ghrelin in normal-cycling female rats and in pregnancy. Obes Res 10:682–687

Guan XM, Yu H, Palyha OC, McKee KK, Feighner SD, Sirinathsinghji DJS, Smith RG, Vanderploeg LHT, Howard AD (1997) Distribution of mRNA encoding the growth hormone secretagogue receptor in brain and peripheral tissues. Mol Brain Res 48:23–29

Holden C (2001) Compulsive behaviors: behavioral addictions: do they exist? Science 294:980–982

Jerlhag E (2008) Systemic administration of ghrelin induces conditioned place preference and stimulates accumbal dopamine. Addict Biol 13:358–363

Jerlhag E, Egecioglu E, Dickson SL, Andersson M, Svensson L, Engel JA (2006) Ghrelin stimulates locomotor activity and accumbal dopamine-overflow via central cholinergic systems in mice: implications for its involvement in brain reward. Addict Biol 11:45–54

Jerlhag E, Egecioglu E, Dickson SL, Douhan A, Svensson L, Engel JA (2007) Ghrelin administration into tegmental areas stimulates locomotor activity and increases extracellular concentration of dopamine in the nucleus accumbens. Addict Biol 12:6–16

Jerlhag E, Egecioglu E, Dickson SL, Engel JA (2010) Ghrelin receptor antagonism attenuates cocaine- and amphetamine-induced locomotor stimulation, accumbal dopamine release, and conditioned place preference. Psychopharmacology 211:415–422

Jerlhag E, Egecioglu E, Dickson SL, Engel JA (2011a) Glutamatergic regulation of ghrelin-induced activation of the mesolimbic dopamine system. Addict Biol 16:82–91

Jerlhag E, Egecioglu E, Dickson SL, Svensson L, Engel JA (2008) Alpha-conotoxin MII-sensitive nicotinic acetylcholine receptors are involved in mediating the ghrelin-induced locomotor stimulation and dopamine overflow in nucleus accumbens. Eur Neuropsychopharmacolology 18:508–518

Jerlhag E, Egecioglu E, Landgren S, Salome N, Heilig M, Moechars D, Datta R, Perrissoud D, Dickson SL, Engel JA (2009) Requirement of central ghrelin signaling for alcohol reward. Proc Natl Acad Sci USA 106:11318–11323

Jerlhag E, Engel JA (2011) Ghrelin receptor antagonism attenuates nicotine-induced locomotor stimulation, accumbal dopamine release and conditioned place preference in mice. Drug Alcohol Depend 117:126–131

Jerlhag E, Janson A-C, Waters S, Engel JA (2012) Concomitant release of ventral tegmental acetylcholine and acumbal dopamine release by ghrelin in rats. Plos One. http://dx.plos.org/10.1371/journal.pone.0049557

Jerlhag E, Landgren S, Egecioglu E, Dickson SL, Engel JA (2011b) The alcohol-induced locomotor stimulation and accumbal dopamine release is suppressed in ghrelin knockout mice. Alcohol 45:341–347

Kaur S, Ryabinin AE (2010) Ghrelin receptor antagonism decreases alcohol consumption and activation of perioculomotor urocortin-containing neurons. Alcohol Clin Exp Res 34:1525–1534

Kojima M, Hosoda H, Date Y, Nakazato M, Matsuo H, Kangawa K (1999) Ghrelin is a growth-hormone-releasing acylated peptide from stomach. Nature 402:656–660

Koob GF, le Moal M (2001) Drug addiction, dysregulation of reward, and allostasis. Neuropsychopharmacology 24:97–129

Kulkosky PJ (1984) Effect of cholecystokinin octapeptide on ethanol intake in the rat. Alcohol 1:125–128

Lanca AJ, Adamson KL, Coen KM, Chow BLC, Corrigall WA (2000) The pedunculopontine tegmental nucleus and the role of cholinergic neurons in nicotine self-administration in the rat: A correlative neuroanatomical and behavioral study. Neuroscience 96:735–742

Landgren S, Engel JA, Andersson ME, Gonzalez-Quintela A, Campos J, Nilsson S, Zetterberg H, Blennow K, Jerlhag E (2009) Association of nAChR gene haplotypes with heavy alcohol use and body mass. Brain Res 1305(Suppl):S72–S79

Landgren S, Engel JA, Hyytia P, Zetterberg H, Blennow K, Jerlhag E (2011a) Expression of the gene encoding the ghrelin receptor in rats selected for differential alcohol preference. Behav Brain Res 221:182–188

Landgren S, Simms JA, Hyytia P, Engel JA, Bartlett SE, Jerlhag E (2012) Ghrelin receptor (GHS-R1A) antagonism suppresses both operant alcohol self-administration and high alcohol consumption in rats. Addict Biol 17:86–94

Landgren S, Simms JA, Thelle DS, Strandhagen E, Bartlett SE, Engel JA, Jerlhag E (2011b) The ghrelin signalling system is involved in the consumption of sweets. PLoS ONE 6:e18170

Larsson A, Edstrom L, Svensson L, Soderpalm B, Engel JA (2005) Voluntary ethanol intake increases extracellular acetylcholine levels in the ventral tegmental area in the rat. Alcohol Alcohol 40:349–358

Larsson A, Engel JA (2004) Neurochemical and behavioral studies on ethanol and nicotine interactions. Neurosci Biobehav Rev 27:713–720

Larsson A, Jerlhag E, Svensson L, Soderpalm B, Engel JA (2004) Is an alpha-conotoxin MII-sensitive mechanism involved in the neurochemical, stimulatory, and rewarding effects of ethanol? Alcohol 34:239–250

Lewis MJ, Johnson DF, Waldman D, Leibowitz SF, Hoebel BG (2004) Galanin microinjection in the third ventricle increases voluntary ethanol intake. Alcohol Clin Exp Res 28:1822–1828

Lof E, Olausson P, Debejczy A, Stomberg R, McIntosh JM, Taylor JR, Soderpalm B (2007) Nicotinic acetylcholine receptors in the ventral tegmental area mediate the dopamine activating and reinforcing properties of ethanol cues. Psychopharmacology 195:333–343

Lyons AM, Lowery EG, Sparta DR, Thiele TE (2008) Effects of food availability and administration of orexigenic and anorectic agents on elevated ethanol drinking associated with drinking in the dark procedures. Alcohol Clin Exp Res 32:1962–1968

Malik S, McGlone F, Bedrossian D, Dagher A (2008) Ghrelin modulates brain activity in areas that control appetitive behavior. Cell Metab 7:400–409

McKee SA, Harrison EL, O'Malley SS, Krishnan-Sarin S, Shi J, Tetrault JM, Picciotto MR, Petrakis IL, Estevez N, Balchunas E (2009) Varenicline reduces alcohol self-administration in heavy-drinking smokers. Biol Psychiatry 66:185–190

Mitchell JM, Teague CH, Kayser AS, Bartlett SE, Fields HL (2012) Varenicline decreases alcohol consumption in heavy-drinking smokers. Psychopharmacology 223:299–306

Morganstern I, Barson JR, Leibowitz SF (2011) Regulation of drug and palatable food overconsumption by similar peptide systems. Curr Drug Abuse Rev 4:163–173

Neugebauer NM, Henehan RM, Hales CA, Picciotto MR (2011) Mice lacking the galanin gene show decreased sensitivity to nicotine conditioned place preference. Pharmacol Biochem Behav 98:87–93

Opland DM, Leinninger GM, Myers MG, Jr (2010) Modulation of the mesolimbic dopamine system by leptin. Brain Res, 1350:65–70

Potenza MN, Steinberg MA, Skudlarski P, Fulbright RK, Lacadie CM, Wilber MK, Rounsaville BJ, Gore MC, Wexler BE (2003) Gambling urges in pathological gambling—a functional magnetic resonance imaging study. Arch Gen Psychiatry 60:828–836

Quarta D, di Francesco C, Melotto S, Mangiarini L, Heidbreder C, Hedou G (2009) Systemic administration of ghrelin increases extracellular dopamine in the shell but not the core subdivision of the nucleus accumbens. Neurochem Int 54:89–94

Rada P, Avena NM, Leibowitz SF, Hoebel BG (2004) Ethanol intake is increased by injection of galanin in the paraventricular nucleus and reduced by a galanin antagonist. Alcohol 33:91–97

Rada PV, Mark GP, Yeomans JJ, Hoebel BG (2000) Acetylcholine release in ventral tegmental area by hypothalamic self-stimulation, eating, and drinking. Pharmacol Biochem Behav 65:375–379

Rasmussen K, Czachura JF, Kallman MJ, Helton DR (1996) The CCK-B antagonist LY288513 blocks the effects of nicotine withdrawal on auditory startle. Neuro Rep 7:1050–1052

Robinson TE, Berridge KC (1993) The neural basis of drug craving—an incentive-sensitization theory of addiction. Brain Res Rev 18:247–291

Salome N, Hansson C, Taube M, Gustafsson-Ericson L, Egecioglu E, Karlsson-Lindahl L, Fehrentz JA, Martinez J, Perrissoud D, Dickson SL (2009) On the central mechanism underlying ghrelin's chronic pro-obesity effects in rats: new insights from studies exploiting a potent ghrelin receptor antagonist. J Neuroendocrinol 21:777–785

Schneider ER, Darby R, Leibowitz SF, Hoebel BG (2007) Orexin, but not ghrelin, injected in the lateral hypothalamus increases alcohol intake in alcohol-drinking rats. Alcohol Clin Exp Res 31:199A–199A

Soderpalm B, Lof E, Ericson M (2009) Mechanistic studies of ethanol's interaction with the mesolimbic dopamine reward system. Pharmacopsychiatry 42(Suppl 1):S87–S94

Suchankova P, Steensland P, Fredriksson I, Engel JA, Jerlhag E (2013) Ghrelin receptor (GHS-R1A) antagonism suppresses both alcohol consumption and the alcohol deprivation effect in rats following long-term voluntary alcohol consumption. Plos One, In press

Steensland P, Simms JA, Holgate J, Richards JK, Bartlett SE (2007) Varenicline, an alpha4beta2 nicotinic acetylcholine receptor partial agonist, selectively decreases ethanol consumption and seeking. Proc Natl Acad Sci U S A 104:12518–12523

Tessari M, Catalano A, Pellitteri M, di Francesco C, Marini F, Gerrard PA, Heidbreder CA, Melotto S (2007) Correlation between serum ghrelin levels and cocaine-seeking behaviour triggered by cocaine-associated conditioned stimuli in rats. Addict Biol 12:22–29

Thiele TE, Navarro M, Sparta DR, Fee JR, Knapp DJ, Cubero I (2003) Alcoholism and obesity: overlapping neuropeptide pathways? Neuropeptides 37:321–337

Thiele TE, Stewart RB, Badia-Elder NE, Geary N, Massi M, Leibowitz SF, Hoebel BG, Egli M (2004) Overlapping peptide control of alcohol self-administration and feeding. Alcohol Clin Exp Res 28:288–294

Tupala E, Tiihonen J (2004) Dopamine and alcoholism: neurobiological basis of ethanol abuse. Prog Neuropsychopharmacol Biol Psychiatry 28:1221–1247

Volkow ND, Fowler JS, Wang GJ (2003a) The addicted human brain: insights from imaging studies. J Clin Inv 111:1444–1451

Volkow ND, Li TK (2004) Drug addiction: the neurobiology of behaviour gone awry. Nat Rev Neurosci 5:963–970

Volkow ND, Wang GJ, Maynard L, Jayne M, Fowler JS, Zhu W, Logan J, Gatley SJ, Ding YS, Wong C, Pappas N (2003b) Brain dopamine is associated with eating behaviors in humans. Int J Eat Disord 33:136–142

von der Goltz C, Koopmann A, Dinter C, Richter A, Rockenbach C, Grosshans M, Nakovics H, Wiedemann K, Mann K, Winterer G, Kiefer F (2010) Orexin and leptin are associated with nicotine craving: a link between smoking, appetite and reward. Psychoneuroendocrinology 35:570–577

Wellman PJ, Clifford PS, Rodriguez J, Hughes S, Eitan S, Brunel L, Fehrentz JA, Martinez J (2011) Pharmacologic antagonism of ghrelin receptors attenuates development of nicotine induced locomotor sensitization in rats. Regul Pept 172:77–80

Wellman PJ, Clifford PS, Rodriguez JA, Hughes S, DI Francesco C, Melotto S, Tessari M, Corsi M, Bifone A, Gozzi A (2012) Brain reinforcement system function is ghrelin dependent: studies in the rat using pharmacological fMRI and intracranial self-stimulation. Addict Biol 17:908–919

Wellman PJ, Davis KW, Nation JR (2005) Augmentation of cocaine hyperactivity in rats by systemic ghrelin. Regul Pept 125:151–154

Wise RA, Bozarth MA (1987) A psychomotor stimulant theory of addiction. Psychol Rev 94:469–492

Wise RA, Rompre PP (1989) Brain dopamine and reward. Annu Rev Psychol 40:191–225

Wren AM, Seal LJ, Cohen MA, Brynes AE, Frost GS, Murphy KG, Dhillo WS, Ghatei MA, Bloom SR (2001a) Ghrelin enhances appetite and increases food intake in humans. J Clin Endocrinol Metab 86:5992–5995

Wren AM, Small CJ, Abbott CR, Dhillo WS, Seal LJ, Cohen MA, Batterham RL, Taheri S, Stanley SA, Ghatei MA, Bloom SR (2001b) Ghrelin causes hyperphagia and obesity in rats. Diabetes 50:2540–2547

Wren AM, Small CJ, Ward HL, Murphy KG, Dakin CL, Taheri S, Kennedy AR, Roberts GH, Morgan DGA, Ghatei MA, Bloom SR (2000) The novel hypothalamic peptide ghrelin stimulates food intake and growth hormone secretion. Endocrinology 141:4325–4328

Yeomans JS, Mathur A, Tampakeras M (1993) Rewarding brain-stimulation—Role of tegmental cholinergic neurons that activate dopamine neurons. Behav Neurosci 107:1077–1087

Zigman JM, Jones JE, Lee CE, Saper CB, Elmquist JK (2006) Expression of ghrelin receptor mRNA in the rat and the mouse brain. J Comp Neurol 494:528–548

Clinical Research on the Ghrelin Axis and Alcohol Consumption

Allison A. Feduccia and Lorenzo Leggio

Abstract Ghrelin, a 28 amino acid orexigenic peptide mainly produced by the stomach, is the endogenous ligand for the growth hormone secretagogue receptor (ghrelin receptor) and regulates a number of physiological processes including energy homeostasis, appetite, gut motility, anxiety, sleep, cardiovascular functions, and inflammation. In addition, preclinical studies demonstrate ghrelin's involvement in reward signaling by its actions in the mesolimbic dopaminergic system, which may enhance the incentive value of food and alcohol rewards. In recent years, clinical studies on alcohol-dependent individuals and healthy controls show that acute and chronic alcohol consumption, as well as abstinence from alcohol, may significantly alter blood ghrelin levels. A positive significant correlation between blood ghrelin levels and alcohol craving has also been reported in alcohol-dependent subjects. Furthermore, single nucleotide polymorphisms (SNPs) within genes encoding ghrelin, i.e., the preproghrelin gene (*GHRL*), and the ghrelin receptor, i.e., growth hormone secretagogue receptor gene (*GHSR*), have been associated with alcohol drinking behaviors and other phenotypic variables related to alcohol dependence; however, these studies need to be replicated in a larger cohort of individuals before strong conclusions can be drawn. In summary, growing, albeit preliminary, human evidence suggests that targeting the ghrelin signaling system may offer a novel pharmacotherapeutic approach for reducing alcohol craving and use in patients with alcohol use disorders.

Keywords Ghrelin · Alcohol dependence · Craving · Relapse · Human · Ghrelin receptor

A. A. Feduccia · L. Leggio (✉)
Section on Clinical Psychoneuroendocrinology and Neuropsychopharmacology, NIAAA and NIDA National Institutes of Health, 10 Center Drive (10CRC/15330) MSC 1108, Room 1-5429, Bethesda, MD 20892-1108, USA
e-mail: lorenzo.leggio@nih.gov

J. Portelli and I. Smolders (eds.), *Central Functions of the Ghrelin Receptor*, The Receptors 25, DOI: 10.1007/978-1-4939-0823-3_8,
© Springer Science+Business Media New York 2014

Introduction: Central Ghrelin Signaling System and the Reward Pathway

After ghrelin was established as a prominent peptide for stimulating hunger and food-reward, it came under investigation in rodent models as a potential mediator of ethanol consumption and in particular, the reinforcement of alcohol drinking. The central ghrelin signaling system contributes significantly to ethanol consumption and drug-seeking behaviors in rodents, as demonstrated by a number of pharmacological and genetic manipulations (see Ghrelin Receptor Antagonism as a Potential Therapeutic Target for Alcohol Use Disorders: A Preclinical Perspective by Jerlhag). Ghrelin interacts with a number of different brain circuits and likely impacts alcohol consumption by a number of mechanisms. To understand the complex, dynamic actions of the ghrelin system on drinking behaviors, one must consider the metabolic roles of ghrelin as well as its direct influence on the reward neural circuitry. This chapter will focus on human studies conducted to investigate the role of ghrelin signaling in alcohol use disorders. Although there is a fairly limited number of clinical reports specifically on ghrelin and alcohol consumption, most do corroborate the theory of the ghrelin system, or disruption of its function, being directly affected by alcohol and contributing to alcohol craving. Human laboratory studies have examined the acute effects of alcohol on blood ghrelin concentrations, as well as the correlation between circulating ghrelin levels and alcohol craving in active drinkers and during abstinence. Investigators have also explored polymorphisms in the genes encoding ghrelin and its receptor for possible associations with substance dependence and drinking behaviors. Furthermore, the potential of the ghrelin signaling system as a pharmacotherapeutic target for the treatment of alcohol use disorders are discussed.

Acute Alcohol Effects on Ghrelin Levels

A few studies on healthy subjects have measured changes in ghrelin levels in response to acute oral alcohol administration. A small initial investigation on alcohol's acute effects on ghrelin levels was conducted in eight healthy individuals who consumed alcohol (0.55 g/kg) during one session and water in a subsequent session. Compared to baseline, serum total ghrelin levels decreased (approximately 13.9 % at 30 min and 17.5 % at 60 min) after consuming the alcoholic drink and exhibited no change after drinking water (Calissendorff et al. 2005). Utilizing the same experimental design, a follow-up study confirmed these results and expanded the findings to demonstrate that both total and octanoylated (active) ghrelin levels continued to decline for more than 5 h after alcohol ingestion (Calissendorff et al. 2006). In another study, nine healthy men consumed 0.6 g/kg alcohol mixed with grapefruit juice (24.8 kj/kg calories) on one day and a matched volume of grapefruit juice (8.17 kj/kg calories) the next day. Total ghrelin levels rapidly declined after drinking the alcoholic beverage, reaching 66 % below baseline after 75 min and remained at this level for the duration of observational period (120 min). Furthermore, ghrelin levels were lower after drinking the alcoholic drink compared to the non-alcoholic juice; however, it is unknown whether this was due to the alcohol or the different caloric content in the drinks (Zimmermann et al. 2007). The authors speculate that ethanol *per se* was responsible for the greater suppression of ghrelin by comparing their results with that of a matched caloric non-alcoholic drink in another study (Callahan et al. 2004). However, this interpretation is confounded since subjects in the latter study had undergone an overnight fast while subjects in the other study had not, which may have pronounced effects on ghrelin's response to caloric intake. Furthermore, the beverages' carbohydrate and lipid content, which is known to affect ghrelin secretion (Overduin et al. 2005), differed between these two studies, thus precluding any comparisons that could show alcohol's inhibition of ghrelin surpasses that of the calories alone. Nonetheless, outcomes across experiments consistently show acute ingestion of alcohol sharply reduces ghrelin levels. However, insufficient evidence exists to conclude whether additional pharmacological effects of alcohol inhibit ghrelin secretion more than its caloric load alone; therefore, further studies are needed to assess the impact of acute alcohol consumption on ghrelin levels.

There are a few proposed mechanisms for alcohol-induced reduction of circulating ghrelin, namely alcohol may control ghrelin secretion by directly inhibiting ghrelin release from gastric mucosa cells and/or indirectly via vagal system activation. Normal plasma ghrelin levels show a cyclical pattern that corresponds to food intake, i.e., gradually rising between meals and sharply decreasing after eating. Therefore, it is reasonable to propose that the caloric content of alcohol consumed acutely may drive this in a similar pattern, and continued consumption of alcohol would prevent the normal escalation of ghrelin observed during fasting.

Role of Ghrelin in Alcohol Craving and Dependence

Clinical studies have investigated fasting ghrelin levels in actively drinking alcohol-dependent individuals as well as during withdrawal and prolonged abstinence from alcohol (Table 1). A small study in 15 actively drinking (i.e., last drink within 24 h) alcohol-dependent males and 15 matched healthy controls showed reduced plasma ghrelin levels in the alcohol-dependent subjects with respect to controls (Addolorato et al. 2006). To further elucidate the role of chronic alcohol consumption on ghrelin levels, Badaoui and colleagues enrolled 51 alcohol-dependent patients (i.e., last drink within 24 h) and 38 healthy social drinkers. On the second day of the study, blood samples were collected from all subjects to determine circulating ghrelin levels and a subgroup of subjects with gastro-esophageal reflux symptoms also received fundic and duodenal biopsies to measure tissue ghrelin

Table 1 Blood ghrelin levels and alcohol craving in active drinkers and abstinent subjects

Subjects[a]	Last drink	Blood ghrelin levels (Alcohol-dependent vs. controls or within-subject)	Correlation between blood ghrelin levels and alcohol craving	Reference
Alcohol-dependent males ($n = 15$) and controls ($n = 15$)	24 h	Decreased (total)	+ correlation between ghrelin levels and craving	(Addolorato et al. 2006)
Alcohol-dependent ($n = 51$) and controls ($n = 38$)	48 h	Decreased (active)[b]	N/A[d]	(Badaoui et al. 2008)
Alcohol-dependent ($n = 24$) and controls ($n = 20$)	24 h 16 days	Decreased No change from baseline	N/A[d]	(de Timary et al. 2012)
Alcohol-dependent [early abstainers ($n = 21$), active drinkers ($n = 97$)] and controls ($n = 24$)	24–72 h Active drinkers	Elevated Elevated and escalated over 7-day abstinence	No correlation between ghrelin levels and craving	(Kraus et al. 2005)
Alcohol-dependent males ($n = 47$) andcontrols ($n = 50$)	>30 days	Elevated[c]	N/A[d]	(Kim et al. 2005)
Alcohol-dependent males ($n = 61$)	14 days	Elevated (active but not total)	+ correlation between ghrelin levels and craving (baseline and after 14-day abstinence)	(Koopmann et al. 2012)
Alcohol-dependent males ($n = 64$)	30 days	Elevated	N/A[d]	(Kim et al. 2013)
Alcohol-dependent ($n = 109$) and controls ($n = 45$)	11 days	Elevated	+ correlation between ghrelin levels and craving (males and females)	(Wurst et al. 2007)
	3 weeks	Normal	+ correlation between ghrelin levels and craving (males only)	

(continued)

Table 1 (continued)

Subjects[a]	Last drink	Blood ghrelin levels(Alcohol-dependent vs. controls or within-subject)	Correlation between blood ghrelin levels and alcohol craving	Reference
Alcohol-dependent subjects ($n = 42$)	72 h	Decreased	+ correlation between baseline ghrelin levels and craving	(Leggio et al. 2012)
	12 weeks	Elevated		

[a] Unless otherwise noted, both genders were included

[b] By contrast, no difference in fundic/duodenal ghrelin mRNA levels was found

[c] There was also a positive correlation between ghrelin levels and duration of abstinence, and a negative correlation between alcohol intake (prior to study) and ghrelin levels during abstinence

[d] Craving was not assessed

content. Alcoholic participants displayed lower active plasma ghrelin levels compared to healthy controls and fundic ghrelin levels trended in the same direction, however, failed to reach significance. Conversely, duodenal ghrelin levels, although not significant, were higher in the alcoholic patients compared to controls, but the authors speculate that since duodenal ghrelin levels are comparatively lower than fundic (in all subjects), the contribution of duodenal ghrelin may be negligible. No differences in ghrelin mRNA levels between groups were found in the fundic and duodenal biopsies, suggesting that alcohol does not affect transcription of ghrelin genes (Badaoui et al. 2008). In support of the studies described so far, a more recent study confirmed that alcohol-dependent subjects ($n = 24$) who drank alcohol until the day of admission had lower fasting ghrelin levels compared to healthy controls ($n = 20$) (de Timary et al. 2012). In contrast, one study found elevated ghrelin levels in both groups (early abstainers and active drinkers) of alcohol-dependent participants compared to non-alcoholic controls (Kraus et al. 2005), although several possible reasons might explain these different findings, e.g., different drinking status (e.g., the latter study included both active drinkers and early abstainers, with the second group having significantly higher ghrelin levels), different ghrelin levels among controls, and different nutritional status. Chronic alcohol use may provide prolonged attenuation of the ghrelin signaling in a manner similar to that seen in obese subjects, essentially desensitizing and deregulating the system. It is possible that the low ghrelin levels parallel nutritional deficiencies associated with chronic alcohol consumption; or ethanol-induced gastritis and atrophic changes (Roberts 1972) may interrupt gastric ghrelin biosynthesis or damage gastric ghrelin-secreting cells directly. Although this specific question is yet to be addressed, long-term alcohol exposure does in fact exacerbate inflammation and damage to mucosa. Interestingly, with a similar consequence, *Helicobacter pylori (H. pylori)* infection induces mucosal impairment and depressed circulating ghrelin (Isomoto et al. 2005). Moreover, after *H. pylori* eradication, plasma ghrelin normalizes within 4 weeks (Nwokolo et al. 2003), which perhaps is a similar adaptation observed after cessation of alcohol use and following weight loss by obese subjects.

Since alterations in ghrelin levels are apparent after long-term alcohol abuse, investigations sought to measure ghrelin level changes during alcohol withdrawal and abstinence. For example, during prolonged alcohol abstinence (>30 days), 47 alcohol-dependent males exhibited enhanced plasma ghrelin levels compared to 50 healthy controls and demonstrated a positive correlation between ghrelin levels and duration of abstinence. In addition, alcohol intake prior to study enrollment was negatively correlated with ghrelin levels during the period of abstinence, indicating that the actual amount of alcohol consumed may directly impact the degree of abstinence-induced ghrelin elevations (Kim et al. 2005). A similar finding was recently reported by Koopmann et al., where active (but not total) ghrelin levels increased in alcohol-dependent persons across a 14-day period of abstinence (Koopmann et al. 2012). In another investigation, alcohol-dependent subjects were divided post-study completion into two groups—alcohol abstainers and non-abstainers—based on whether or not they voluntarily consumed alcohol during the 12-week study. Results showed that abstainers had lower baseline ghrelin levels compared to non-abstainers; furthermore, across the 12-week duration of the trial, ghrelin levels increased in the abstainers group and decreased in the non-abstainers (Leggio et al. 2012). In a similar manner, during early withdrawal from alcohol (24–72 h) alcohol-dependent individuals ($n = 21$) had higher ghrelin levels than active drinkers ($n = 97$) and controls ($n = 24$), and during a 7-day withdrawal period ghrelin levels escalated in the "active drinker group" (Kraus et al. 2005). In support, a within-subject comparison of plasma ghrelin levels in alcohol-dependent participants revealed enhanced levels after 30 days alcohol-free compared to prior measurements taken when subjects were still consuming alcohol (Kim et al. 2013). In another study, after approximately 11 days of alcohol withdrawal, patients had significantly higher ghrelin levels than controls; however, in this study ghrelin levels decreased across the 3-week rehabilitation period and differences were no longer detectable at the end of the study (Wurst et al. 2007). de Timary et al. report significantly lower baseline ghrelin levels in actively drinking alcohol-dependent subjects enrolled in an inpatient study compared to controls; however, levels remained constant across the 16-day withdrawal period (de Timary et al. 2012).

Since ghrelin levels appear to be altered after chronic alcohol use and during various periods of alcohol withdrawal, researchers sought to understand if these changes may correlate with craving for alcohol. During a 12-week study in alcohol-dependent subjects, two assessments of craving, Penn Alcohol Craving Scale (PACS) and the Obsessive Compulsive Drinking Scale (OCDS), and plasma ghrelin levels were collected at four time points. Baseline ghrelin levels were positively correlated with PACS scores at the 2-week time point and with both craving measures at 6 and 12 weeks; by contrast, post-baseline ghrelin levels at various intervals (2-week, 6-week, and 12-week time points) were not associated with craving (Leggio et al. 2012). A positive association of ghrelin and alcohol craving measured with the OCDS was also demonstrated in a study of 15 male actively drinking, alcohol-dependent individuals (Addolorato et al. 2006). Consistent with these findings, Koopmann et al. also observed a positive correlation between OCDS scores and active (but not total) ghrelin levels in alcohol-dependent males ($n = 61$) on the first day of admission and after a 14-day period of abstinence (Koopmann et al. 2012). In a different study, determination of craving by use of the OCDS, Alcohol Urge Questionnaire (AUQ), and Alcohol Visual Analog Scale (A-VAS) and blood ghrelin levels were attained from 45 healthy controls and 109 alcohol-dependent patients withdrawn from alcohol for approximately 11 days at the time of the first assessments and again at the end of a 3-week rehabilitation program. Authors reported gender-specific differences in ghrelin levels in the alcohol-dependent group, such that female patients had higher ghrelin levels than males at both time points. For all patients, ghrelin levels correlated with AUQ craving scores at the first time point but when genders were analyzed separately association of specific craving scales with ghrelin levels differed between sexes at week 0. Furthermore, at week 3, correlations between ghrelin levels and alcohol craving disappeared with the exception of a correlation in only males' ghrelin levels with one question on the OCDS measure (Wurst et al. 2007). It should be noted, however, that analyses of single items of the OCDS questionnaire are not typically reported in the literature. On the other hand, Kraus et al. did not detect a relationship between ghrelin and craving scales in alcohol-dependent participants (Kraus et al. 2005).

Taken together, findings reported on ghrelin levels in alcohol-dependent individuals, excluding those that investigated the acute effects of alcohol on ghrelin release, are somewhat inconsistent and many factors likely account for the divergent findings, such as gender, time since last use of alcohol, calculation/measurement of ghrelin (active vs. total, ghrelin/BMI), and the number of subjects enrolled. Future studies should aim for more controlled experiments to better understand the direct effects of ghrelin on craving for alcohol and rigorously attempt to account for potentially confounding variables such as BMI and diet.

In conclusion, the overall hypothesis driven by the human studies conducted to date is that, in alcoholic individuals, the ghrelin system might play a role in alcohol craving and consumption; however, at this time, it is unclear whether the ghrelin-mediated effects are due simply to its appetitive properties and hedonic effects, or as the animal literature suggests (see Ghrelin Receptor Antagonism as a Potential Therapeutic Target for Alcohol Use Disorders: A Preclinical Perspective by Jerlhag), the ghrelin system has pronounced and direct activity on reward processing. Therefore, more human evidence is needed to clearly characterize ghrelin's role in alcohol craving and central activation of reward-related pathways.

Gastric Bypass Surgery: Disrupting Gut Hormone Activation of the Reward Circuit

Gastric bypass surgery has provided the opportunity to investigate the physiological effects and behavioral consequences of dramatically blunting the ghrelin signaling system. While the procedure itself reduces food intake by gastric restriction, long-term maintenance of reduced body weight is also attributed to suppression of gut-derived molecules, such as ghrelin, that play a major regulatory role in consumption. After gastric bypass surgery, as expected, individuals exhibit a marked reduction (77 %) in circulating ghrelin levels compared to controls and show almost a complete loss of the normal occurring fluctuations in ghrelin levels between meals (Cummings et al. 2002; Morinigo et al. 2004). For some time it was debatable if these alterations in ghrelin were due to post-surgery weight loss or to changes incurred at the gastric level. To answer this question, a study compared ghrelin levels after significant weight loss between a diet modification alone group and a gastric bypass surgery group. Since diet-induced weight loss alone did not have significant effects on ghrelin levels, it appears post-surgery weight loss was not responsible for the alterations in ghrelin level but were likely due to changes incurred by gastric ghrelin-secreting cells (Cummings et al. 2002).

Individuals with current alcohol use disorders are often excluded from receiving gastric bypass. Therefore, it has become of interest to understand if depressed ghrelin signaling may affect alcohol drinking and craving. A percentage of individuals increase alcohol consumption after gastric bypass surgery and have a higher incidence of dependence (Buffington 2007; Conason et al. 2013; Svensson et al. 2013) with one study reporting as high as 28.4 % of the sample indicated difficulties controlling alcohol intake after surgery while only 4.5 % did before bariatric surgery (Kalarchian et al. 2007). In contrast, other subjects reported drinking considerably less and some patients with previous alcohol dependence abstained after surgery (Buffington 2007; King et al. 2012). Reviews of the literature suggest that the risk of alcohol use disorder post-surgery is extremely low with the exception that people with a prior diagnosis of alcohol dependence had an increased incidence of relapse (Suzuki et al. 2012; Buffington 2007). Interestingly, the metabolism of alcohol is altered after gastric bypass surgery, with breath alcohol levels reaching greater levels and longer alcohol clearance times compared to controls, which could potentially impact the risk for alcohol-related problems (Kalarchian et al. 2002; Hagedorn et al. 2007). Consistent with this finding, 84 % of individuals who have undergone bariatric surgery report experiencing intoxication after consuming a small amount of alcohol and 29 % indicated prolonged intoxication after alcohol consumption compared to their presurgery experiences (Kalarchian et al. 2007; Ertelt et al. 2008). At this time, insufficient data is available to determine whether gastric bypass surgery alters the risk of alcohol dependence since subjects report both increased and decreased alcohol intake after surgery and importantly, the role of ghrelin in post-surgery consumption of alcohol is yet to be determined. Additionally, results from rodent experiments are also bidirectional—increased and decreased ethanol drinking—following gastric bypass surgery (Davis et al. 2012, 2013; Hajnal et al. 2012, 2013), indicating multiple factors influence drinking behaviors post-surgery and likely exceed that of alterations in ghrelin secretion alone.

Association of Ghrelin/Ghrelin Receptor Polymorphisms with Heavy Alcohol Use

Given ghrelin's widespread effects on various systems and functions of the body, many studies have investigated the genetic variability of the ghrelin system and its relationship with BMI, type 2 diabetes, eating disorders, cardiovascular disease, and more recently alcohol consumption (Table 2). A haplotype analysis of 10 SNPs in genes encoding preproghrelin (*GHRL*) and the ghrelin receptor (*GHRS*) was conducted in a sample of Spanish subjects ($n = 417$) which were divided into cohorts of non-drinkers, moderate, and heavy alcohol drinkers. One SNP in the *GHSR* gene, rs2232165, was associated with alcohol consumption. Another SNP located in *GHSR* that has previously been associated with obesity and bulimia,

Table 2 Significant associations between ghrelin signaling system SNPs and substance abuse traits

GHRL	GHRS	Traits	# SNPs tested	Subjects[a]	Reference
Haplotypes	rs2948694 Haplotypes	Increased weight/BMI in heavy drinkers	10	Non-drinkers, moderate, and heavy drinkers ($n = 417$)	(Landgren et al. 2008)
	rs2232165	Alcohol consumption			
Haplotypes		Withdrawal symptoms and incidence of paternal alcohol dependence	10	Alcohol-dependent females ($n = 113$) and controls ($n = 212$)	(Landgren et al. 2010)
rs42451 rs35680		Self-transcendence	10	Type 1 alcohol-dependent ($n = 84$) and controls ($n = 32$)	(Landgren et al. 2011)
	rs495225	Decreased self-directedness			
	rs495225 rs2948694	Novelty seeking	10	Non-dependent ($n = 317$)	(Hansson et al. 2012)
rs34911341 rs696217		No significant findings	2	Alcohol-dependent males ($n = 70$) and controls ($n = 68$)	(Leggio et al. 2012)
rs49684677	rs2948694	Amphetamine addiction severity score More prevalent in amphetamine-dependent individuals	10	Amphetamine-dependent ($n = 104$) and controls ($n = 310$)	(Suchankova et al. 2013)
rs696217		Higher depression and anxiety scores	1	Methamphetamine-dependent ($n = 118$) and controls ($n = 144$)	(Yoon et al. 2005)

[a] Unless otherwise noted, both genders were included

Abbreviations: GHRL pro-ghrelin gene; *GHSR* growth hormone secretagogue receptor gene; *SNP* single nucleotide polymorphisms; *BMI* body mass index rs2948694, was associated with increased weight and BMI in heavy drinkers in this population as were haplotypes in both *GHRL* and *GHRS* (Landgren et al. 2008). In a second study by Landgren et al., the same 10 tag SNPs were investigated in alcohol-dependent ($n = 113$) and social drinking ($n = 212$) Swedish females. Results failed to confirm an increased risk for alcohol dependence as in the prior study but revealed weak associations of two similar *GHRL* haplotypes in the alcohol-dependent group with withdrawal symptoms and incidence of paternal alcohol dependence (Landgren et al. 2010). In a small sample of type 1 alcoholics ($n = 84$) and healthy controls ($n = 32$), the 10 SNPs in *GHRL* and *GHSR* were analyzed to determine associations with risk for alcohol dependence or factors of the Temperament and Character Inventory. While no SNPs were associated with alcohol dependence, one SNP in *GHRS* was associated with decreased self-directedness and two SNPs in GHRL were associated with measures of self-transcendence in alcohol-dependent subjects (Landgren et al. 2011). A different study also utilizing the same personality inventory found significant associations of two *GHSR* SNPs with novelty seeking; moreover, one of these *GHRS* SNPs, rs495225, was the same as that in the before-mentioned study although associated with a different trait measure (Hansson et al. 2012). In a recent report, GHRL and *GHSR* SNPs were analyzed in Swedish amphetamine-dependent subjects ($n = 104$) and healthy controls ($n = 310$) and findings indicated a *GHSR* SNP rs2948694 was more prevalent in amphetamine-dependent individuals and the *GHRL* SNP rs4684677 was significantly associated with Addiction Severity Interview scores (Suchankova et al. 2013). A common SNP in *GHRL* (rs696217) was explored in a Korean sample to compare allelic frequencies of methamphetamine-dependent subjects ($n = 118$) and controls ($n = 144$). No significant differences were found between groups for this specific polymorphism; however, the methamphetamine-dependent participants carrying the Met72 allele had higher depression and anxiety scores compared to Leu homozygote subjects (Yoon et al. 2005). In another study, frequency comparisons of two polymorphisms of *GHRL*, rs34911341 and rs696217, in a small sample of alcohol-dependent ($n = 70$) and control subjects ($n = 68$) revealed no significant differences nor did analysis of alcohol-dependent carriers of the Leu72Met variant with drinks/day, age of onset, years of addiction, or family history of alcoholism (Leggio et al. 2012).

Together, findings suggest risk for drug dependence and personality traits commonly expressed in people with alcohol use disorders might be influenced by genetic variations in the ghrelin signaling system. However, strong conclusions from these candidate gene studies should be heeded. Due to the nature of candidate gene approaches and the analysis of small sample sizes within limited populations, findings from these studies hold the potential of generating false-positive results and/or findings that might not be replicable in different populations. As such, larger case-controlled studies are needed to validate and replicate these preliminary findings; furthermore, the functional significance of these SNPs remains to be elucidated. To date, no genome-wide studies report positive findings for ghrelin polymorphisms and alcohol/drug dependence.

Therapeutic Target for the Treatment of Alcoholism

Alcoholism remains a prominent public health concern; however to date, there are only a few approved medications with suboptimal effectiveness for the treatment of alcohol use disorders. Overall, preclinical and clinical evidence indicate that the ghrelin signaling system plays a pivotal role in alcohol craving and likely contributes to alcohol use and relapse. Since elevated ghrelin levels arise during alcohol cessation and correlate with a high incidence of craving, a reasonable hypothesis, as supported by pharmacological and genetic manipulations in rodents, is that disrupting or lowering ghrelin signaling during alcohol cessation may diminish craving and decrease alcohol consumption. Future studies should aim to address the therapeutic utility of blockade of central ghrelin receptors or reducing peripheral secretion of ghrelin from the gut in alcohol-dependent individuals. At this time, it is unknown if ghrelin receptor 1a antagonism in humans would result in a global suppression of reward (i.e., food or other abused drugs), or possibly more specific to alcohol reward, and if targeting peripheral ghrelin secretion may potentially limit side effects of central ghrelin receptor 1a antagonists. If outcomes from rodent experiments translate to humans, then targeting the ghrelin signaling pathways may offer a novel approach for the treatment of alcohol use disorders.

Conclusions

In the past decade, a significant amount of research has been underway to elucidate the role of the ghrelin signaling system in appetitive processes and specifically, reinforcement of food and, more recently, alcohol reward. Ongoing accumulation of data suggests that ghrelin levels are significantly impacted by both acute and chronic alcohol consumption and alterations in this system following long-term consumption of alcohol may drive relapse in alcohol-dependent individuals due to enhanced craving for alcohol during withdrawal and abstinence. In addition, genetic variants within the ghrelin system are associated with various aspects of dependence, but larger studies are needed to validate these findings and determine the functional role of these polymorphisms. Taken together, available evidence warrants further investigations into the ghrelin system as a potential target for the treatment for alcohol dependence.

Acknowledgments This work was supported by the NIH Intramural Research Programs of the National Institute on Alcohol Abuse and Alcoholism (NIAAA) and the National Institute on Drug Abuse (NIDA).

References

Addolorato G, Capristo E, Leggio L, Ferrulli A, Abenavoli L, Malandrino N, Farnetti S, Domenicali M, D'Angelo C, Vonghia L, Mirijello A, Cardone S, Gasbarrini G (2006) Relationship between ghrelin levels, alcohol craving, and nutritional status in current alcoholic patients. Alcohol Clin Exp Res 30(11):1933–1937

Badaoui A, De Saeger C, Duchemin J, Gihousse D, de Timary P, Starkel P (2008) Alcohol dependence is associated with reduced plasma and fundic ghrelin levels. Eur J Clin Invest 38(6):397–403

Buffington CK (2007) Alcohol use and health risks: survey results. Bariatric Times 4:21–23

Calissendorff J, Danielsson O, Brismar K, Rojdmark S (2005) Inhibitory effect of alcohol on ghrelin secretion in normal man. Eur J Endocrinol / Eur Federation Endocrine Soc 152(5):743–747

Calissendorff J, Danielsson O, Brismar K, Rojdmark S (2006) Alcohol ingestion does not affect serum levels of peptide YY but decreases both total and octanoylated ghrelin levels in healthy subjects. Metab, Clin Exp 55(12):1625–1629

Callahan HS, Cummings DE, Pepe MS, Breen PA, Matthys CC, Weigle DS (2004) Postprandial suppression of plasma ghrelin level is proportional to ingested caloric load but does not predict intermeal interval in humans. J Clin Endocrinol Metab 89(3):1319–1324

Conason A, Teixeira J, Hsu CH, Puma L, Knafo D, Geliebter A (2013) Substance use following bariatric weight loss surgery. JAMA Surg 148(2):145–150

Cummings DE, Weigle DS, Frayo RS, Breen PA, Ma MK, Dellinger EP, Purnell JQ (2002) Plasma ghrelin levels after diet-induced weight loss or gastric bypass surgery. N Engl J Med 346 (21):1623–1630

Davis JF, Schurdak JD, Magrisso IJ, Mul JD, Grayson BE, Pfluger PT, Tschöp MH, Seeley RJ, Benoit SC (2012) Gastric bypass surgery attenuates ethanol consumption in ethanol-preferring rats. Biol Psychiatry 72(5):354–360

Davis JF, Tracy AL, Schurdak JD, Magrisso IJ, Grayson BE, Seeley RJ, Benoit SC (2013) Roux en Y Gastric Bypass Increases Ethanol Intake in the Rat. Obes Surg

de Timary P, Cani PD, Duchemin J, Neyrinck AM, Gihousse D, Laterre P-F, Badaoui A, Leclercq S, Delzenne NM, Stärkel P (2012) The loss of metabolic control on alcohol drinking in heavy drinking alcohol-dependent subjects. PLoS ONE 7(7)

Ertelt TW, Mitchell JE, Lancaster K, Crosby RD, Steffen KJ, Marino JM (2008) Alcohol abuse and dependence before and after bariatric surgery: a review of the literature and report of a new data set. Surg Obes Relat Dis: Off J Am Soc Bariat Surg 4(5):647–650

Hagedorn JC, Encarnacion B, Brat GA, Morton JM (2007) Does gastric bypass alter alcohol metabolism? Surg Obes Relat Dis: Off J Am Soc Bariatr Surg 3(5):543–548; discussion 548

Hajnal A, Thanos PK, Volkow ND (2013) Notes on "Roux en Y gastric bypass increases ethanol intake in the rat by Davis et al. Obes Surg

Hajnal A, Zharikov A, Polston JE, Fields MR, Tomasko J, Rogers AM, Volkow ND, Thanos PK (2012) Alcohol reward is increased after Roux-en-Y gastric bypass in dietary obese rats with differential effects following Ghrelin Antagonism. PLoS ONE 7(11)

Hansson C, Shirazi RH, Naslund J, Vogel H, Neuber C, Holm G, Anckarsater H, Dickson SL, Eriksson E, Skibicka KP (2012) Ghrelin influences novelty seeking behavior in rodents and men. PLoS ONE 7(12):e50409

Isomoto H, Ueno H, Saenko VA, Mondal MS, Nishi Y, Kawano N, Ohnita K, Mizuta Y, Ohtsuru A, Yamashita S, Nakazato M, Kohno S (2005) Impact of *Helicobacter pylori* infection on gastric and plasma ghrelin dynamics in humans. Am J Gastroenterol 100(8):1711–1720

Kalarchian MA, Marcus MD, Levine MD, Courcoulas AP, Pilkonis PA, Ringham RM, Soulakova JN, Weissfeld LA, Rofey DL (2007) Psychiatric disorders among bariatric surgery candidates: relationship to obesity and functional health status. Am J Psychiatry 164(2):328–334; quiz 374

Kalarchian MA, Marcus MD, Wilson GT, Labouvie EW, Brolin RE, LaMarca LB (2002) Binge eating among gastric bypass patients at long-term follow-up. Obes Surg 12(2):270–275

Kim DJ, Yoon SJ, Choi B, Kim TS, Woo YS, Kim W, Myrick H, Peterson BS, Choi YB, Kim YK, Jeong J (2005) Increased fasting plasma ghrelin levels during alcohol abstinence. Alcohol Alcohol 40(1):76–79

Kim JH, Kim SJ, Lee WY, Cheon YH, Lee SS, Ju A, Kim DJ (2013) The effects of alcohol abstinence on BDNF, ghrelin, and leptin secretions in alcohol-dependent patients with glucose intolerance. Alcohol Clin Exp Res 37 Suppl 1:E52–58

King WC, Chen J-Y, Mitchell JE, Kalarchian MA, Steffen KJ, Engel SG, Courcoulas AP, Pories WJ, Yanovski SZ (2012) Prevalence of alcohol use disorders before and after bariatric surgery. JAMA 307(23):2516–2525

Koopmann A, von der Goltz C, Grosshans M, Dinter C, Vitale M, Wiedemann K, Kiefer F (2012) The association of the appetitive peptide acetylated ghrelin with alcohol craving in early abstinent alcohol dependent individuals. Psychoneuroendocrinology 37(7):980–986

Kraus T, Schanze A, Gröschl M, Bayerlein K, Hillemacher T, Reulbach U, Kornhuber J, Bleich S (2005) Ghrelin levels are increased in alcoholism. Alcohol Clin Exp Res 29(12):2154–2157

Landgren S, Berglund K, Jerlhag E, Fahlke C, Balldin J, Berggren U, Zetterberg H, Blennow K, Engel JA (2011) Reward-related genes and personality traits in alcohol-dependent individuals: a pilot case control study. Neuropsychobiology 64(1):38–46

Landgren S, Jerlhag E, Hallman J, Oreland L, Lissner L, Strandhagen E, Thelle DS, Zetterberg H, Blennow K, Engel JA (2010) Genetic variation of the ghrelin signaling system in females with severe alcohol dependence. Alcohol Clin Exp Res 34(9):1519–1524

Landgren S, Jerlhag E, Zetterberg H, Gonzalez-Quintela A, Campos J, Olofsson U, Nilsson S, Blennow K, Engel JA (2008) Association of pro-ghrelin and GHS-R1A gene polymorphisms and haplotypes with heavy alcohol use and body mass. Alcohol Clin Exp Res 32(12):2054–2061

Leggio L, Ferrulli A, Cardone S, Nesci A, Miceli A, Malandrino N, Capristo E, Canestrelli B, Monteleone P, Kenna GA, Swift RM, Addolorato G (2012) Ghrelin system in alcohol-dependent subjects: role of plasma ghrelin levels in alcohol drinking and craving. Addict Biol 17(2):452–464

Morinigo R, Casamitjana R, Moize V, Lacy AM, Delgado S, Gomis R, Vidal J (2004) Short-term effects of gastric bypass surgery on circulating ghrelin levels. Obes Res 12(7):1108–1116

Nwokolo CU, Freshwater DA, O'Hare P, Randeva HS (2003) Plasma ghrelin following cure of *Helicobacter pylori*. Gut 52(5):637–640

Overduin J, Frayo RS, Grill HJ, Kaplan JM, Cummings DE (2005) Role of the duodenum and macronutrient type in ghrelin regulation. Endocrinology 146(2):845–850

Roberts DM (1972) Chronic gastritis, alcohol, and non-ulcer dyspepsia. Gut 13(10):768–774

Suchankova P, Jerlhag E, Jayaram-Lindstrom N, Nilsson S, Toren K, Rosengren A, Engel JA, Franck J (2013) Genetic variation of the ghrelin signalling system in individuals with amphetamine dependence. PLoS ONE 8(4):e61242

Suzuki J, Haimovici F, Chang G (2012) Alcohol use disorders after bariatric surgery. Obes Surg 22(2):201–207

Svensson PA, Anveden A, Romeo S, Peltonen M, Ahlin S, Burza MA, Carlsson B, Jacobson P, Lindroos AK, Lonroth MDH, Maglio C, Naslund MDI, Sjoholm K, Wedel Ph DH, Soderpalm MDB, Sjostrom L, Carlsson LM (2013) Alcohol consumption and alcohol problems after bariatric surgery in the Swedish Obese Subjects (SOS) study. Obesity

Wurst FM, Graf I, Ehrenthal HD, Klein S, Backhaus J, Blank S, Graf M, Pridzun L, Wiesbeck GA, Junghanns K (2007) Gender differences for ghrelin levels in alcohol-dependent patients and differences between alcoholics and healthy controls. Alcohol Clin Exp Res 31(12):2006–2011

Yoon SJ, Pae CU, Lee H, Choi B, Kim TS, Lyoo IK, Kwon DH, Kim DJ (2005) Ghrelin precursor gene polymorphism and methamphetamine dependence in the Korean population. Neurosci Res 53(4):391–395
Zimmermann US, Buchmann A, Steffin B, Dieterle C, Uhr M (2007) Alcohol administration acutely inhibits ghrelin secretion in an experiment involving psychosocial stress. Addict Biol 12(1):17–21

Part IV
Ghrelin Plays a Role in Various Physiological and Pathophysiological Brain Functions

Ghrelin and Sleep Regulation

Éva Szentirmai and Levente Kapás

Abstract Classic models of sleep regulation posit that the timing and amount of sleep are determined by the duration of prior wakefulness and whether or not the circadian phase is favorable for sleep. Growing body of evidence indicates, however, that in addition to these factors, other signals from the external and internal environment also play a key role in sleep regulation. Changes in metabolic environment, such as positive and negative energy balance, adiposity, postprandial state, and shifts in lipolytic activity all have fundamental effects on sleep. The signaling mechanisms that connect metabolism to sleep regulation include hormones of the gastrointestinal tract and the adipose tissue. Ghrelin signaling in the brain has emerged as one of the key components of the arousal system that is activated in negative energy states and possibly under other physiological conditions. We review recent human and animals studies on the role of ghrelin in sleep regulation and in the function of biological clocks.

Keywords Growth hormone secretagogue · Metabolism · Sleep · Ghrelin · Food-entrainable oscillator · Food anticipatory activity · Thermoregulation

Introduction

The relationship between sleep and metabolism has long been recognized. Cross-species correlational studies in mammals revealed an interaction between daily sleep amounts and resting metabolic rate (Zepelin and Rechtschaffen 1974; Allison

É. Szentirmai (✉) · L. Kapás
Washington, Wyoming, Alaska, Montana and Idaho (WWAMI) Medical Education Program
and Department of Integrative Physiology and Neuroscience, Sleep and Performance
Research Center, Washington State University, PO Box 1495 Spokane,
WA 99210-1495, USA
e-mail: eszentirmai@wsu.edu

J. Portelli and I. Smolders (eds.), *Central Functions of the Ghrelin Receptor*,
The Receptors 25, DOI: 10.1007/978-1-4939-0823-3_9,

and Cicchetti 1976). The sleep-related decrease in body temperature and energy expenditure led to the speculation that the primordial function of sleep is energy conservation (Allison and van Twyver 1970). Though the function of sleep is yet to be determined, we have gained a more complex understanding of the relationship between sleep and metabolism. It is not only sleep that is accompanied by secondary changes in metabolism but it also became evident that shifts in metabolism and the overall nutritional state of the organism lead to adaptive responses in sleep. The fundamental nature of metabolic signals in regulating vigilance was demonstrated in decorticated rats over 40 years ago. Decorticated rats that are almost constantly active and show no periods of inactivity longer than 12 min appear to sleep for 2 h after tube feeding; this is the only time they ever appear to sleep (Sorenson and Ellison 1970). Subsequent studies revealed that increased caloric intake (Jacobs and McGinty 1971; Borbély 1977; Danguir et al. 1979; Danguir 1987; Hansen et al. 1998), high adiposity (Guan et al. 2008) as well as increased lipolysis (Danguir and Nicolaidis 1980) facilitate sleep, whereas negative energy states promote arousal (Borbély 1977; Danguir and Nicolaidis 1979; Yamanaka et al. 2003; Gelegen et al. 2006; Kanizsai et al. 2009; Esposito et al. 2012) in intact rats and mice.

Several hypothalamic areas, such as the suprachiasmatic nucleus, lateral hypothalamus (LH), and ventromedial hypothalamic nucleus are implicated in the regulation of both sleep and metabolism/food intake (Berthoud 2002). These structures express receptors for multiple hormones of the gastrointestinal tract and adipose tissue. These hormones can modulate sleep and metabolism independent of one another and recent evidence indicates that they play a role in aligning vigilance with the current metabolic state of the body. Several gastrointestinal hormones, such as cholecystokinin (Shemyakin and Kapás 2001) and adipokines, e.g., leptin (Sinton et al. 1999), facilitate sleep in positive energy states. The main physiological stimulus for ghrelin secretion from the gastrointestinal tract and for the activation of the brain ghrelin system is fasting. In this chapter we review the evidence that ghrelin signaling in the brain is a key component of the arousal system which facilitates wakefulness in negative energy states.

Ghrelin and Sleep

The Effects of Ghrelin and Growth Hormone Secretagogues on Sleep in Humans

Ghrelin is the endogenous ligand for the growth hormone secretagogue receptor 1a (GHS-R1a; hereafter we refer to it as the ghrelin receptor). Growth hormone secretagogues (GHSs) are synthetic peptides with the ability to bind to pituitary membranes and stimulate growth hormone (GH) secretion (reviewed in Cruz and Smith 2008). Based on the close relationship between hormones of the somatotropic axis and sleep regulation (reviewed Obál and Krueger 2004), a series of human studies was initiated at the Max-Plank Institute in Munich and at the Université Libre de Bruxelles in the 1990s to test GHSs for their potential use in sleep medicine.

Unfortunately, an unambiguous picture regarding the effects of GHSs on human sleep did not emerge from these experiments. Different GHSs showed divergent and often opposite effects on sleep and the effects of a given GHS varied according to the route of administration and gender. For example, pulsatile intravenous (i.v.) administration of growth hormone-releasing peptide (GHRP)-6 at doses that stimulate nocturnal GH and cortisol release did not affect total sleep time, sleep latency, or other sleep-related electroencephalographic (EEG) variables, only a modest increase in stage 2 sleep occurred (Frieboes et al. 1995). Oral administration of GHRP-6 before bedtime significantly decreased total sleep, increased sleep latency, and suppressed stage 2 non-rapid-eye-movement sleep (NREMS), while the sublingual or intranasal administration of the peptide had no effect (Frieboes et al. 1999). Another orally active GHS, MK-677, increased stage 4 NREMS by 50 % and rapid-eye-movement sleep (REMS) by 20 % in young male subjects (Copinschi et al. 1997). In subsequent experiments, the same research group could not detect any effect on sleep when they used a different GHS, GHRP-2 (Moreno-Reyes et al. 1998); this led them to conclude that the sleep-promoting actions of MK-677 are likely independent of the activation of ghrelin receptors. Finally, the most potent GHS agonist, hexarelin, decreased deep, stage 4 sleep, and suppressed EEG delta power during NREMS (also called slow-wave activity of the EEG, or SWA, a measure used for characterizing the intensity/depth of NREMS) (Frieboes et al. 2004).

After the identification of ghrelin as the endogenous GHS receptor agonist, the focus from GHSs shifted to the endogenous ligand. In an elegant series of experiments at the Max-Plank Institute, Munich, ghrelin was tested in young and elder men and women using various administration schedules. In these studies, sleep recordings were complemented with simultaneous measurements of plasma GH and cortisol levels.

In the first study using ghrelin, repeated i.v. bolus injections of the ghrelin to young healthy males enhanced NREMS (particularly stage 4) and increased EEG delta activity. REMS and measures of sleep continuity and sleep architecture remained unchanged but ghrelin stimulated both cortisol and GH secretion (Weikel et al. 2003). Since components of both the somatotropic and the hypothalamic-pituitary-adrenal axes are implicated in sleep regulation (reviewed in Steiger 2007) the interactions of ghrelin with growth hormone-releasing hormone (GHRH) and corticotropin-releasing hormone (CRH) were also investigated. Placebo, ghrelin alone or in combination with CRH or GHRH was injected during the first part of the night to young men. In contrast to the previous findings, ghrelin itself had no effect on any of the sleep parameters during the first half of the night, while stage 2 NREMS slightly increased during the second half of the night. Co-administration of GHRH or CRH with ghrelin did not modify the sleep effects of ghrelin but potentiated its GH- and cortisol-stimulating effects, respectively (Kluge et al. 2008).

The timing of treatment and gender are two major factors that appear to determine the effects of ghrelin on human sleep. When sequential injections of ghrelin are performed in the early morning hours instead of the first half of the night, ghrelin loses its effects on sleep and EEG while the GH- and cortisol-stimulating effects persist (Kluge et al. 2007a). Ghrelin does not have any effect on sleep in young (Kluge et al. 2007b) or postmenopausal women (Kluge et al. 2010), while it still stimulates plasma GH and cortisol levels. Recently, ghrelin was tested in patients with major depression (Kluge et al. 2011). In depressed men, ghrelin significantly reduced the time spent awake in the second part of the night without affecting any parameters of NREMS and REMS. In depressed women, however, waking was not altered but the amount of REMS was decreased.

The emerging picture from the human experiments is that when ghrelin is administered in i.v. boluses during the first part of the night it induces slight increases in sleep in young men. In healthy women, young or elder, ghrelin has no effects on sleep parameters. Further, if ghrelin is administered during the second part of the night, it loses its modest sleep-promoting activity even in males. Ghrelin consistently stimulates plasma GH and cortisol levels in each age group and gender irrespective of the timing of administration.

The Effects of Ghrelin on Sleep in Rats and Mice

While human studies did not result in a clear picture of the role of ghrelin in sleep regulation and the potential benefit of GHSs in sleep medicine, findings from animal studies suggest that central ghrelin signaling is a key component of the arousal system. This notion is supported by three sets of findings. One, ghrelin neurons and ghrelin receptors are strategically located in hypothalamic circuits of central arousal mechanisms. Two, central administration of exogenous ghrelin increases wakefulness in rats and mice. Three, deficiency of ghrelin signaling impairs the function of wake-inducing mechanisms.

The Effects of Systemic Ghrelin Treatment on Sleep

The first animal study to investigate the effects of ghrelin on sleep used multiple bolus i.v. injections in rats (Tolle et al. 2002). The first injection was performed 1.5 h before the end of the light period and two subsequent treatments were given during the early dark phase. Ghrelin increased wakefulness and decreased NREMS and REMS for 30 min after all three injections. In mice, intraperitoneal injection of a single dose of ghrelin, 400 ug/kg, caused a transient increase in sleep in wild-type (WT) animals but had no effect in mice with nonfunctional GHRH receptors (Obál et al. 2003).

A comprehensive, third study was recently published on the effects of systemic ghrelin treatment in mice (Szentirmai 2012). Ghrelin was administered in a wide dose-range, during both the dark and light periods and sleep recordings were complemented by food intake, body temperature, and motor activity measurements. Systemic administration of ghrelin did not induce changes in sleep, motor activity, or body temperature. Food intake was significantly increased indicating that physiologically relevant doses of ghrelin were tested. As discussed below, the activation of central ghrelin-sensitive mechanisms has profound wake-promoting effects. The findings with systemic ghrelin injections suggest that activation of those ghrelin receptors which are directly accessible for circulating ghrelin does not activate the same central wake-promoting mechanisms. Considerable evidence suggests that circulating ghrelin acts on peripheral targets to stimulate feeding (Date et al. 2002, 2006; Asakawa et al. 2001). The finding that activation of these peripheral mechanisms has no effect on sleep-wake activity indicates that the wake-promoting and feeding-stimulating actions of ghrelin are independent of each other.

The Effects of Central Ghrelin Treatment on Sleep

Ghrelin is a gut-brain peptide present in the gastrointestinal tract and in neurons of the central nervous system. In the brain, ghrelin is produced by neurons of the hypothalamus, mainly in the arcuate nucleus (ARC), LH, paraventricular nucleus (PVN), and by a population of neurons in the hypothalamic area adjacent to the third ventricle. Ghrelin-positive axon terminals are present in the LH, ARC, and PVN where they synapse with orexinergic and neuropeptide Y (NPY)-producing neurons (reviewed in Kageyama et al. 2010). The ghrelin receptor is expressed in hypothalamic nuclei including the ARC, suprachiasmatic nucleus, LH, and ventromedial hypothalamic nucleus (Guan et al. 1997; Mitchell et al. 2001; Zigman et al. 2006) and also in the hippocampus and nodes of the mesolimbic reward system, such as the ventral tegmental area and nucleus accumbens (Abizaid et al. 2006; Skibicka et al. 2011).

Since most of the structures that express ghrelin receptors play a key role in modulating vigilance, the possible role of central ghrelin signaling in sleep-wake regulation received considerable attention. In the first study on the central ghrelin system in sleep regulation, the effects of intracerebroventricular (i.c.v.) bolus injection of ghrelin on sleep, feeding, and behavior in rats were investigated (Szentirmai et al. 2006). I.c.v. administration of ghrelin induced significant dose-dependent increases in wakefulness with the concomitant suppression of both NREMS and REMS. Increased wakefulness was accompanied by signs of behavioral activation in the first hour including increased locomotor activity, eating, drinking, grooming, and exploration. The first feeding bout occurred 10 min after the injection and eating continued throughout the first hour of the light period. Feeding behavior per se, however, was not responsible for the wake-promoting effect of ghrelin since it remained present when animals did not have access to food. Similar to rats, i.c.v. injection of ghrelin induced robust sleep-suppressing effects in mice with the suppression of EEG SWA (Szentirmai 2012). Consistent with its wake-promoting effects, i.c.v. (Carlini et al. 2002; Jerlhag et al. 2006; Jászberényi et al. 2006), intra-ventral tegmental area, or intra-laterodorsal tegmental area (Jerlhag et al. 2007) injections of ghrelin increase locomotor activity.

To identify the central target(s) for the wake-inducing effects of ghrelin, a series of microinjection studies were performed in rats. The LH was a likely target since it has a central role in the regulation of vigilance (McGinty and Szymusiak 2003) and feeding (Bernardis and Bellinger 1996), and ghrelin axon terminals (Mitchell et al. 2001; Toshinai et al. 2003; Cowley et al. 2003) as well as ghrelin receptors are present (Mitchell et al. 2001; Harrold et al. 2008). Ghrelin microinjections into the LH induced a dose-dependent increase in wakefulness and feeding and the suppression of sleep (Szentirmai et al. 2007a). Similar to the effects of ghrelin, intra-LH microinjections of NPY (Szentirmai and Krueger 2006a) and i.c.v. administration of orexin also induced wakefulness (Hagan et al. 1999) and eating (Sakurai et al. 1998). Based on these findings, a hypothesis was proposed that increased wakefulness and feeding are two parallel outputs of a hypothalamic ghrelin-sensitive circuitry that also involves NPY and orexin neurons. The hypothalamic orexin-ghrelin-NPY circuit integrates metabolic, circadian, and possibly homeostatic sleep signals as well as signals arising from the external environment. The activation of this mechanism triggers a coordinated behavioral sequence characteristic of transiently occurring negative energy states such as the first hours of the activity period in rats (dark onset syndrome) (Szentirmai et al. 2007a). Orexinergic projections from the LH are likely to constitute the major output of this circuit. Ghrelin-containing axon-terminals make direct synaptic connections with orexin neurons (Toshinai et al. 2003) and i.c.v. or local micro-injection of ghrelin activates orexin cells (Lawrence et al. 2002; Toshinai et al. 2003; Olszewski et al. 2003; Yamanaka et al. 2003). Increased feeding in response to ghrelin application into the LH is, at least in part, mediated by orexin (Toshinai et al. 2003). Since orexin is a key component in the arousal system (reviewed in (Sakurai et al. 2010), it is possible that orexin mechanisms also play a role in ghrelin-induced arousal responses. In addition to the LH, other potential wake-inducing targets for ghrelin include the medial preoptic area and the PVN. Ghrelin microinjection into these nuclei also facilitates wakefulness (Szentirmai et al. 2007a) and feeding (Wren et al. 2001; Szentirmai et al. 2007a) in rats. The importance of the medial preoptic area in the hypothalamic sleep-regulating system (McGinty and Szymusiak 2003) and the PVN in arousal, autonomic, and behavioral responses to stressors (Pfaff et al. 2005) is well-documented.

In addition to ghrelin, the preproghrelin gene (Ppg) also codes for obestatin (Zhang et al. 2005; Seim et al. 2011) and for other alternative mRNA transcripts with unidentified corresponding peptide products (Seim et al. 2007). Although the physiological role of obestatin is still unclear, some evidence suggests that its effects on feeding may be the opposite of ghrelin's (Zhang et al. 2005; Bresciani et al. 2006; Lagaud et al. 2007; Carlini et al. 2007). Such dichotomy is also apparent in the effects of the two peptides on sleep. While ghrelin has strong wake-promoting activities, obestatin induces significant increases in NREMS and shortens sleep latency (Szentirmai and Krueger 2006b).

Sleep in Preproghrelin and Ghrelin Receptor Knockout (KO) Mice

Sleep has been studied in mice with the congenital deletion of the ghrelin receptor and mice with the deletion of the Ppg gene. The major difference in the two KO models is that while ghrelin signaling is absent in both genotypes, in the Ppg KO animals, not only is ghrelin affected but also all other products of the Ppg gene, e.g., obestatin. Despite the potent food intake-stimulating effect of exogenously administered ghrelin, deletion of the Ppg gene does not affect normal phenotype, body weight, growth rate, body composition, and food intake in mice (Sun et al. 2003). While exogenous administration of ghrelin causes robust increases in wakefulness, Ppg KO animals (Szentirmai et al. 2007b) and ghrelin receptor deficient mice (Esposito et al. 2012) do not show fundamental sleep-wake deficiencies under standard laboratory conditions other than increased sleep fragmentation. Interestingly, other transgenic mouse strains lacking key components of arousal-promoting system such as orexin (Mochizuki et al. 2004) or histamine (Parmentier et al. 2002) also show relatively normal sleep duration yet fragmented sleep architecture. Sleep deprivation in Ppg KO (Szentirmai et al. 2007b) and ghrelin receptor KO (Esposito et al. 2012) mice induces normal rebound sleep suggesting that KO animals retain adequate homeostatic sleep-promoting mechanisms.

In general, the lack of a major change in spontaneous sleep-wake activity under standard laboratory conditions in KO animals does not necessarily reflect the significance of the affected signaling system in sleep regulation. Genetic lesions of other major arousal mechanisms, such as the serotonin, histamine, norepinephrine, CRH, and orexin systems, also fail to elicit gross changes in spontaneous sleep-wake activity (Chemelli et al. 1999; Boutrel et al. 2002; Parmentier et al. 2002; Hunsley and Palmiter 2003; Romanowski et al. 2010). This is probably due to the redundancy within the arousal system and/or to the evolvement of compensatory mechanisms during development. Redundancy may not be sufficient for the more robust activation of the arousal system under natural conditions, when the animal is confronted with environmental challenges such as a change in the environment or food shortage. In fact, when ghrelin receptor KO mice are subjected to cage change or short-term fasting, they fail to mount adequate arousal responses normally seen in WT animals (Esposito et al. 2012). This indicates that intact ghrelin signaling is required for the normal function of fundamental wake-promoting mechanisms in mice and supports the hypothesis that ghrelin signaling plays a key role in the arousal system.

When challenged with subthermoneutral temperature, Ppg KO mice show increased cold sensitivity manifested as significantly reduced body temperature and suppressed sleep compared to WTs. When WT mice are fasted in the cold, they exhibit short hypothermic bouts accompanied by increased sleep. In Ppg KO mice, however, fasting in cold exacerbates their thermoregulatory deficiency, their body temperature drops precipitously reaching near-ambient temperature and EEG-defined sleep disappears (Szentirmai et al. 2009). Interestingly, ghrelin receptor KO mice show no such sensitivity to the combination of cold and fasting suggesting that the thermoregulatory and sleep deficits in Ppg KO mice are not due to the lack of ghrelin signaling. Replacement of obestatin, the other major Ppg gene product, by using osmotic minipumps partially rescues the phenotype suggesting that the lack of obestatin in Ppg KO mice may, at least in part, be responsible for the observed thermoregulatory and sleep deficit (Szentirmai et al. 2009).

Ghrelin and the Biological Clocks

Diurnal changes in sleep-wake activity, feeding, and metabolism are driven by biological clock(s). It has been proposed that ghrelin plays a role in the function of these clocks. The spontaneous, free running, rhythms of biological clocks are slightly different from 24 h, they need to be synchronized (entrained) to the 24 h solar day by photic or metabolic stimuli. The "master" clock in the suprachiasmatic nucleus is entrained by light whereas the other major biological clock, the food-entrainable oscillator (FEO) is entrained by periodic feeding. The location and the molecular machinery of the FEO are unknown; its existence is inferred from the manifestations of its activity, such as the food anticipatory activity (FAA). FAA is characterized by increased behavioral activity, elevated corticosterone secretion, and rises in body temperature 1–4 h before scheduled feeding time when feeding is restricted to a few hours daily. Signaling from the gastrointestinal system is key to the activity of the FEO. The FEO itself could be located in the intestinal system; in this case, gut-to-brain signaling is required as an output signal from the clock to the brain to elicit the characteristic behavioral and autonomic responses. If the FEO is centrally located then gut-derived signals are required to serve as input signals for the entrainment of the clock. Gastrointestinal hormones, the secretions of which are phase-locked to feeding, are likely candidates to serve as such a signal. Ghrelin emerged as an obvious candidate that may integrate feeding- and metabolism-related signals directed to the FEO or ghrelin signaling may be part of the molecular machinery in FEO itself. The secretion of ghrelin is locked to feeding activity; ghrelin plasma levels are elevated during fasting and suppressed after eating (Tschöp et al. 2000; Cummings et al. 2001; Bodosi et al. 2004). In scheduled feeding paradigms, plasma ghrelin levels increase in parallel with FAA (Drazen et al. 2006).

The role of ghrelin in the FEO was investigated in five independent studies by using ghrelin KO or ghrelin receptor KO transgenic mice. In three studies using ghrelin receptor KO animals, food-anticipatory motor activity was measured either as wheel running activity or spontaneous locomotion, or both (Blum et al. 2009;

LeSauter et al. 2009; Davis et al. 2011). In two of the experiments (Blum et al. 2009; LeSauter et al. 2009), clear entrainment to scheduled feeding developed in KO mice indicating the retained integrity of the FEO function. In one experiment, ghrelin receptor KO animals increased their activity closer to the expected feeding time, which could be interpreted as a more efficient and improved clock function (LeSauter et al. 2009). In the other two experiments (Blum et al. 2009; Davis et al. 2011), the intensity of the anticipatory response, as measured by the number of wheel rotations or spontaneous activity counts, was attenuated but still present. This likely reflects a change in the activities of the effector mechanisms driven by the FEO, not an impaired clock function. In a comprehensive fourth study, Ppg KO mice were used and three parameters of FAA were measured simultaneously. Both normal and Ppg KO mice developed normal food anticipatory responses manifested as increases in waking time, motor activity, and body temperature. Neither the timing nor the intensity of the FAA responses was different between the two genotypes (Szentirmai et al. 2010). In the most recent study, FAA was measured by an automated behavior recognition system in Ppg KO and WT mice. After 2 weeks of a restricted feeding paradigm, both genotypes exhibited robust anticipatory behavior (Gunapala et al. 2011). The findings that clear entrainment persists to restricted feeding in both ghrelin receptor and Ppg KO mice indicates that the time-keeping function of FEO does not require intact ghrelin signaling.

Conclusion

A growing body of evidence indicates that ghrelin is a member of the group of neuropeptides/hormones that play a role in sleep regulation. Central ghrelinergic mechanisms—as part of the hypothalamic ghrelin-orexin-NPY circuit—are posited to play a role in promoting wakefulness and feeding. The assumed function of the circuit is to integrate metabolic and circadian signals and set sleep-wake activity according to the metabolic needs of the organism. For example, in negative energy states, such as fasting, the activity of the circuit facilitates arousal thus sets vigilance to a state that is favorable for replenishing energy stores, e.g., foraging.

While data from animal studies favor this hypothesis, human data are less clear and often contradictory. Due to the cost and the inherent complexity of human sleep experiments, crucial studies, such as establishing dose-response relationships for ghrelin, are lacking. Furthermore, human studies only investigate the function of the circulating ghrelin pool and give only limited information about central ghrelinergic mechanisms.

References

Abizaid A, Liu ZW, Andrews ZB et al (2006) Ghrelin modulates the activity and synaptic input organization of midbrain dopamine neurons while promoting appetite. J Clin Invest 116:3229–3239

Allison T, Van Twyver H (1970) The evolution of sleep. Nat Hist 79:55–65

Allison T, Cicchetti DV (1976) Sleep in mammals: ecological and constitutional correlates. Science 194:732–734

Asakawa A, Inui A, Kaga T et al (2001) Ghrelin is an appetite-stimulatory signal from stomach with structural resemblance to motilin. Gastroenterology 120:337–345

Bernardis LL, Bellinger LL (1996) The lateral hypothalamic area revisited: ingestive behavior. Neurosci Biobehav Rev 20:189–287

Berthoud HR (2002) Multiple neural systems controlling food intake and body weight. Neurosci Biobehav Rev 26:393–428

Blum ID, Patterson Z, Khazall R et al (2009) Reduced anticipatory locomotor responses to scheduled meals in ghrelin receptor deficient mice. Neuroscience 164:351–359

Bodosi B, Gardi J, Hajdu I et al (2004) Rhythms of ghrelin, leptin, and sleep in rats: effects of the normal diurnal cycle, restricted feeding, and sleep deprivation. Am J Physiol Regul Integr Comp Physiol 287:R1071–R1079

Borbély AA (1977) Sleep in the rat during food deprivation and subsequent restitution of food. Brain Res 124:457–471

Boutrel B, Monaca C, Hen R et al (2002) Involvement of 5-HT1A receptors in homeostatic and stress-induced adaptive regulations of paradoxical sleep: studies in 5-HT1A knock-out mice. J Neurosci 22:4686–4692

Bresciani E, Rapetti D, Dona F et al (2006) Obestatin inhibits feeding but does not modulate GH and corticosterone secretion in the rat. J Endocrinol Invest 29:RC16–RC18

Carlini VP, Monzon ME, Varas MM et al (2002) Ghrelin increases anxiety-like behavior and memory retention in rats. Biochem Biophys Res Commun 299:739–743

Carlini VP, Schioth HB, Debarioglio SR (2007) Obestatin improves memory performance and causes anxiolytic effects in rats. Biochem Biophys Res Commun 352:907–912

Chemelli RM, Willie JT, Sinton CM et al (1999) Narcolepsy in orexin knockout mice: molecular genetics of sleep regulation. Cell 98:437–451

Copinschi G, Leproult R, Van Onderbergen A et al (1997) Prolonged oral treatment with MK-677, a novel growth hormone secretagogue, improves sleep quality in man. Neuroendocrinology 66:278–286

Cowley MA, Smith RG, Diano S et al (2003) The distribution and mechanism of action of ghrelin in the CNS demonstrates a novel hypothalamic circuit regulating energy homeostasis. Neuron 37:649–661

Cruz CR, Smith RG (2008) The growth hormone secretagogue receptor. Vitam Horm 77:47–88

Cummings DE, Purnell JQ, Frayo RS et al (2001) A preprandial rise in plasma ghrelin levels suggests a role in meal initiation in humans. Diabetes 50:1714–1719

Danguir J (1987) Cafeteria diet promotes sleep in rats. Appetite 8:49–53

Danguir J, Nicolaidis S (1979) Dependence of sleep on nutrients' availability. Physiol Behav 22:735–740

Danguir J, Nicolaidis S (1980) Circadian sleep and feeding patterns in the rat: possible dependence on lipogenesis and lipolysis. Am J Physiol 238:E223–E230

Danguir J, Nicolaidis S, Gerard H (1979) Relations between feeding and sleep patterns in the rat. J Comp Physiol Psychol 93:820–830

Date Y, Murakami N, Toshinai K et al (2002) The role of the gastric afferent vagal nerve in ghrelin-induced feeding and growth hormone secretion in rats. Gastroenterology 123:1120–1128

Date Y, Shimbara T, Koda S et al (2006) Peripheral ghrelin transmits orexigenic signals through the noradrenergic pathway from the hindbrain to the hypothalamus. Cell Metab 4:323–331

Davis JF, Choi DL, Clegg DJ et al (2011) Signaling through the ghrelin receptor modulates hippocampal function and meal anticipation in mice. Physiol Behav 103:39–43

Drazen DL, Vahl TP, D'Alessio DA et al (2006) Effects of a fixed meal pattern on ghrelin secretion: evidence for a learned response independent of nutrient status. Endocrinology 147:23–30

Esposito M, Pellinen J, Kapás L et al (2012) Impaired wake-promoting mechanisms in ghrelin receptor-deficient mice. Eur J Neurosci 35:233–243

Frieboes RM, Murck H, Maier P et al (1995) Growth hormone-releasing peptide-6 stimulates sleep, growth hormone, ACTH and cortisol release in normal man. Neuroendocrinology 61:584–589

Frieboes RM, Murck H, Antonijevic IA et al (1999) Effects of growth hormone-releasing peptide-6 on the nocturnal secretion of GH, ACTH and cortisol and on the sleep EEG in man: role of routes of administration. J Neuroendocrinol 11:473–478

Frieboes RM, Antonijevic IA, Held K et al (2004) Hexarelin decreases slow-wave sleep and stimulates the secretion of GH, ACTH, cortisol and prolactin during sleep in healthy volunteers. Psychoneuroendocrinology 29:851–860

Gelegen C, Collier DA, Campbell IC et al (2006) Behavioral, physiological, and molecular differences in response to dietary restriction in three inbred mouse strains. Am J Physiol Endocrinol Metab 291:E574–E581

Guan XM, Yu H, Palyha OC et al (1997) Distribution of mRNA encoding the growth hormone secretagogue receptor in brain and peripheral tissues. Brain Res Mol Brain Res 48:23–29

Guan Z, Vgontzas AN, Bixler EO et al (2008) Sleep is increased by weight gain and decreased by weight loss in mice. Sleep 31:627–633

Gunapala KM, Gallardo CM, Hsu CT et al (2011) Single gene deletions of orexin, leptin, neuropeptide Y, and ghrelin do not appreciably alter food anticipatory activity in mice. PLoS One 6:e18377

Hagan JJ, Leslie RA, Patel S et al (1999) Orexin A activates locus coeruleus cell firing and increases arousal in the rat. Proc Natl Acad Sci USA 96:10911–10916

Hansen MK, Kapás L, Fang J et al (1998) Cafeteria diet-induced sleep is blocked by subdiaphragmatic vagotomy in rats. Am J Physiol 274:R168–R174

Harrold JA, Dovey T, Cai XJ et al (2008) Autoradiographic analysis of ghrelin receptors in the rat hypothalamus. Brain Res 1196:59–64

Hunsley MS, Palmiter RD (2003) Norepinephrine-deficient mice exhibit normal sleep-wake states but have shorter sleep latency after mild stress and low doses of amphetamine. Sleep 26:521–526

Jacobs BL, McGinthy DJ (1971) Effects of food deprivation on sleep and wakefulness in the rat. Exp Neurol 30:212–222

Jászberényi M, Bujdosó E, Bagosi Z et al (2006) Mediation of the behavioral, endocrine and thermoregulatory actions of ghrelin. Horm Behav 50:266–273

Jerlhag E, Egecioglu E, Dickson SL et al (2006) Ghrelin stimulates locomotor activity and accumbal dopamine-overflow via central cholinergic systems in mice: implications for its involvement in brain reward. Addict Biol 11:45–54

Jerlhag E, Egecioglu E, Dickson SL et al (2007) Ghrelin administration into tegmental areas stimulates locomotor activity and increases extracellular concentration of dopamine in the nucleus accumbens. Addict Biol 12:6–16

Kageyama H, Takenoya F, Shiba K et al (2010) Neuronal circuits involving ghrelin in the hypothalamus-mediated regulation of feeding. Neuropeptides 44:133–138

Kanizsai P, Garami A, Solymár M et al (2009) Energetics of fasting heterothermia in TRPV1-KO and wild type mice. Physiol Behav 96:149–154

Kluge M, Schussler P, Zuber V et al (2007a) Ghrelin administered in the early morning increases secretion of cortisol and growth hormone without affecting sleep. Psychoneuroendocrinology 32:287–292

Kluge M, Schussler P, Zuber V et al (2007b) Ghrelin enhances the nocturnal secretion of cortisol and growth hormone in young females without influencing sleep. Psychoneuroendocrinology 32:1079–1085

Kluge M, Schussler P, Bleninger P et al (2008) Ghrelin alone or co-administered with GHRH or CRH increases non-REM sleep and decreases REM sleep in young males. Psychoneuroendocrinology 33:497–506

Kluge M, Gazea M, Schussler P et al (2010) Ghrelin increases slow wave sleep and stage 2 sleep and decreases stage 1 sleep and REM sleep in elderly men but does not affect sleep in elderly women. Psychoneuroendocrinology 35:297–304

Kluge M, Schussler P, Dresler M et al (2011) Effects of ghrelin on psychopathology, sleep and secretion of cortisol and growth hormone in patients with major depression. J Psychiatr Res 45:421–426

Lagaud GJ, Young A, Acena A et al (2007) Obestatin reduces food intake and suppresses body weight gain in rodents. Biochem Biophys Res Commun 2007 357:264–269

Lawrence CB, Snape AC, Baudoin FM et al (2002) Acute central ghrelin and GH secretagogues induce feeding and activate brain appetite centers. Endocrinology 143:155–162

LeSauter J, Hoque N, Weintraub M et al (2009) Stomach ghrelin-secreting cells as food-entrainable circadian clocks. Proc Natl Acad Sci USA 2009 106:13582–13587

McGinty D, Szymusiak R (2003) Hypothalamic regulation of sleep and arousal. Front Biosci 8:s1074–s1083

Mitchell V, Bouret S, Beauvillain JC et al (2001) Comparative distribution of mRNA encoding the growth hormone secretagogue-receptor (GHS-R) in Microcebus murinus (Primate, lemurian) and rat forebrain and pituitary. J Comp Neurol 429:469–489

Mochizuki T, Crocker A, McCormack S et al (2004) Behavioral state instability in orexin knock-out mice. J Neurosci 2004 24:6291–6300

Moreno-Reyes R, Kerkhofs M, L'Hermite-Baleriaux M et al (1998) Evidence against a role for the growth hormone-releasing peptide axis in human slow-wave sleep regulation. Am J Physiol 274:E779–E784

Obal F Jr, Krueger JM (2004) GHRH and sleep. Sleep Med Rev 8:367–377

Obal F Jr, Alt J, Taishi P et al (2003) Sleep in mice with nonfunctional growth hormone-releasing hormone receptors. Am J Physiol Regul Integr Comp Physiol 284:R131–R139

Olszewski PK, Li D, Grace MK et al (2003) Neural basis of orexigenic effects of ghrelin acting within lateral hypothalamus. Peptides 24:597–602

Parmentier R, Ohtsu H, Djebbara-Hannas Z et al (2002) Anatomical, physiological, and pharmacological characteristics of histidine decarboxylase knock-out mice: evidence for the role of brain histamine in behavioral and sleep-wake control. J Neurosci 22:7695–7711

Pfaff D, Westberg L, Kow LM (2005) Generalized arousal of mammalian central nervous system. J Comp Neurol 493:86–91

Romanowski CP, Fenzl T, Flachskamm C et al (2010) Central deficiency of corticotropin-releasing hormone receptor type 1 (CRH-R1) abolishes effects of CRH on NREM but not on REM sleep in mice. Sleep 33:427–436

Sakurai T, Amemiya A, Ishii M et al (1998) Orexins and orexin receptors: a family of hypothalamic neuropeptides and G protein-coupled receptors that regulate feeding behavior. Cell 92:573–585

Sakurai T, Mieda M, Tsujino N (2010) The orexin system: roles in sleep/wake regulation. Ann N Y Acad Sci 1200:149–161

Seim I, Collet CC, Herington AC et al (2007) Revised genomic structure of the human ghrelin gene and identification of novel exons, alternative splice variants and natural antisense transcripts. BMC Genomics 8:298

Seim I, Josh P, Cunningham P et al (2011) Ghrelin axis genes, peptides and receptors: recent findings and future challenges. Mol Cell Endocrinol 340:3–9

Shemyakin A, Kapás L (2001) L-364, 718, a cholecystokinin-A receptor antagonist, suppresses feeding-induced sleep in rats. Am J Physiol Regul Integr Comp Physiol 280:R1420–R1426

Sinton CM, Fitch TE, Gershenfeld HK (1999) The effects of leptin on REM sleep and slow wave delta in rats are reversed by food deprivation. J Sleep Res 8:197–203

Skibicka KP, Hansson C, Alvarez-Crespo M et al (2011) Ghrelin directly targets the ventral tegmental area to increase food motivation. Neuroscience 180:129–137

Sorenson CA, Ellison GD (1970) Striatal organization of feeding behavior in the decorticate rat. Exp Neurol 29:162–174

Steiger A (2007) Neurochemical regulation of sleep. J Psychiatr Res 41:537–552

Sun Y, Ahmed S, Smith RG (2003) Deletion of ghrelin impairs neither growth nor appetite. Mol Cell Biol 23:7973–7981

Szentirmai E (2012) Central but not systemic administration of ghrelin induces wakefulness in mice. PLoS One 7:e41172

Szentirmai E, Krueger JM (2006a) Central administration of neuropeptide Y induces wakefulness in rats. Am J Physiol Regul Integr Comp Physiol 291:R473–R480

Szentirmai E, Krueger JM (2006b) Obestatin alters sleep in rats. Neurosci Lett 404:222–226

Szentirmai E, Hajdu I, Obal F et al (2006) Ghrelin-induced sleep responses in ad libitum fed and food-restricted rats. Brain Res 1088:131–140

Szentirmai É, Kapás L, Krueger JM (2007a) Ghrelin microinjection into forebrain sites induces wakefulness and feeding in rats. Am J Physiol Regul Integr Comp Physiol 292:R575–R585

Szentirmai É, Kapás L, Sun Y et al (2007b) Spontaneous sleep and homeostatic sleep regulation in ghrelin knockout mice. Am J Physiol Regul Integr Comp Physiol 293:R510–R517

Szentirmai É, Kapás L, Sun Y et al (2009) The preproghrelin gene is required for the normal integration of thermoregulation and sleep in mice. Proc Natl Acad Sci USA 106:14069–14074

Szentirmai É, Kapás L, Sun Y, Smith RG, Krueger JM et al (2010) Restricted feeding-induced sleep, activity, and body temperature changes in normal and preproghrelin-deficient mice. Am J Physiol Regul Integr Comp Physiol 298:R467–R477

Tolle V, Bassant MH, Zizzari P et al (2002) Ultradian rhythmicity of ghrelin secretion in relation with GH, feeding behavior, and sleep-wake patterns in rats. Endocrinology 143:1353–1361

Toshinai K, Date Y, Murakami N et al (2003) Ghrelin-induced food intake is mediated via the orexin pathway. Endocrinology 144:1506–1512

Tschöp M, Smiley DL, Heiman ML (2000) Ghrelin induces adiposity in rodents. Nature 407:908–913

Weikel JC, Wichniak A, Ising M et al (2003) Ghrelin promotes slow-wave sleep in humans. Am J Physiol Endocrinol Metab 284:E407–E415

Wren AM, Small CJ, Abbott CR et al (2001) Ghrelin causes hyperphagia and obesity in rats. Diabetes 50:2540–2547

Yamanaka A, Beuckmann CT, Willie JT et al (2003) Hypothalamic orexin neurons regulate arousal according to energy balance in mice. Neuron 38:701–713

Zepelin H, Rechtschaffen A (1974) Mammalian sleep, longevity, and energy metabolism. Brain Behav Evol 10:425–470

Zhang JV, Ren PG, Avsian-Kretchmer O et al (2005) Obestatin, a peptide encoded by the ghrelin gene, opposes ghrelin's effects on food intake. Science 310:996–999

Zigman JM, Jones JE, Lee CE et al (2006) Expression of ghrelin receptor mRNA in the rat and the mouse brain. J Comp Neurol 494:528–548

Ghrelin and Memory

Nicolas Kunath and Martin Dresler

Abstract The 28-amino acid peptide ghrelin was originally identified as an orexigenic hormone involved in the regulation of an organism's energy homeostasis. Besides its role in metabolic processes, accumulating evidence suggests that ghrelin also plays an important role in the cognitive aspects of energy homeostasis, in particular learning and memory. Several studies in rodents confirm enhancing effects of ghrelin on fear learning, object recognition and spatial memory, in particular when given before the encoding phase of memory formation. Several mechanisms of action, intracellular signaling pathways and neurotransmitters involved in ghrelin's effects on memory processes have been revealed, including serotonin, dopamine, neuropeptide Y, and nitric oxide, whose interplay affects hippocampal processes of neuroplasticity. Research on the role of ghrelin in the cognition of nonrodent species including humans is sparse and less conclusive, sometimes even suggesting memory-impairing effects of ghrelin. However, the increasing body of evidence demonstrating memory-supporting and neuroprotective effects in rodent models calls for further research that elucidates ghrelin's effects on human cognition and its prospect in the therapy and prophylaxis of neurological diseases.

Keywords Ghrelin · Learning · Memory · Consolidation · Cognition · Plasticity · Enhancement

N. Kunath (✉) · M. Dresler
Max Planck Institute of Psychiatry, Kraepelinstraße 2-10 80804 Munich, Germany
e-mail: nicolas_kunath@mpipsykl.mpg.de

M. Dresler
e-mail: dresler@mpipsykl.mpg.de

J. Portelli and I. Smolders (eds.), *Central Functions of the Ghrelin Receptor*,
The Receptors 25, DOI: 10.1007/978-1-4939-0823-3_10,
© Springer Science+Business Media New York 2014

Introduction

Imagine a squirrel collecting food for its winter stocks, gathering nuts and acorns, hiding them in a forest's soil close to roots, under bushes, and beneath the bark of a tree. After several weeks, temperatures get low, snow falls and the squirrel's energy resources hit the bottom line. However, it somehow remembers where the food it once gathered is hidden, with its memory as the only key to survival. What is the underlying mechanism of this impressive feat? Or, looking beyond squirrels, how did our early ancestors remember the place where they last found delicious blueberries, where they last slaughtered a mammoth? Organisms depending on external sources of energy to survive somehow have to develop efficient ways to engrave into their mind the places where food awaits them. Besides its ambivalent role in the regulation of hunger and satiety, accumulating evidence suggests that ghrelin might play an important role in the cognitive aspects of energy homeostasis.

Ghrelin and Memory: Behavioral Data

Memory is not a uniform phenomenon, but can be divided into different phases and subsystems. A growing number of studies have demonstrated that ghrelin influences several aspects of learning and memory formation. The influence of ghrelin on memory was first observed not in squirrels, but in a rat model. While basal memory processes like habituation to a novel environment were not affected by ghrelin administration, fear learning was; in a step-down inhibitory avoidance task, ghrelin given intracerebroventricularly (i.c.v.) significantly increased memory retention in a dose-dependent manner (Carlini et al. 2002). Similarly, i.c.v. ghrelin dose-dependently enhanced memory in a passive avoidance task (Goshadrou et al. 2013). A second type of memory influenced by ghrelin is object recognition: ghrelin administration to the hippocampus improved this kind of memory in rats (Carlini et al. 2008), an effect that could also be shown for non-peptide ghrelin receptor agonists (Atcha et al. 2009). Likewise in mice, decreases in object recognition performance due to chronic food restriction were counteracted by ghrelin administration (Carlini et al. 2008). Furthermore, spatial memory retention was shown to be enhanced by subcutaneous injections of ghrelin or the ghrelin mimetic LY444711 in mice (Diano et al. 2006), and by oral or subcutaneous administration of the non-peptide ghrelin receptor agonists GSK894490A and CP-464709-18 in rats (Atcha et al. 2009). Interestingly, ghrelin receptor deficient mice express impairments in spatial learning, but not avoidance learning compared to wild types (Diano et al. 2006). Recently, the role of ghrelin has been investigated for different phases of memory formation; while ghrelin improved memory retention when administered before training, it had no significant effect on memory performance if administered before retrieval (Carlini et al. 2010a).

Compared to the accumulating evidence for the relationship between ghrelin and memory in mice and rats, little is known about ghrelin's role in human cognition. Some studies observed a negative correlation between ghrelin and memory. In a group of non-demented elderly adults, ghrelin was shown to have a negative effect on declarative memory (Spitznagel et al. 2010). In another study, although results failed to be statistically significant, nocturnal administration of ghrelin seemed to have impairing effects on sleep-related neuroplasticity: gains in a sequential motor skill task, normally seen after a night of sleep were leveled by ghrelin (Dresler et al. 2010). In contrast, results suggesting a role for ghrelin as a neuroprotective agent in Alzheimer's, Parkinson's disease, and ischemia have been reported (Gahete et al. 2010, 2011; Theodoropoulou et al. 2012; Kenny et al. 2013; Bayliss und Andrews 2013; Dos Santos et al. 2013) (see also Chaps. 12 and 13).

It is possible that ghrelin's effects on memory differ across species. In a study conducted with neonatal chicks, the administration of ghrelin led to an impairment of memory retention (Carvajal et al. 2009). However, also in rodents, for which memory enhancing effects of ghrelin were most consistently shown, conflicting findings have been observed. In a more recent study, growth hormone secretagogue receptor (ghrelin receptor) 1a knockout-mice habituated faster to a novel environment and performed better in a Morris water maze task than controls, suggesting a controversial role of the ghrelin receptor 1a in memory processing (Albarran-Zeckler et al. 2012). Of the two identified ghrelin receptor subtypes, only ghrelin receptor 1a binds active acyl ghrelin whereas ghrelin receptor 1b seems to have modulating effects on the 1a-type (Laviano et al. 2012). Clearly, further studies are needed to elucidate ghrelin's functions in memory processing across different species, and in particular in human cognition.

Neural Structures

The hippocampus is the central structure of interest for memory processes (Turner 1969). Traditionally, synaptic connections within the hippocampus are described with the concept of a trisynaptic loop (Neves et al. 2008). Inputs from the perforant path arrive at the dentate gyrus which projects to the CA3 subfield, itself projecting via Schaffer collaterals to the CA1 subfield whose projections exit the hippocampus via subiculum and fornix. Ghrelin enters into this synaptic loop as shown by the discovery of a significant change of long-term potentiation (LTP) in hippocampal slices induced by ghrelin (Diano et al. 2006)—LTP is seen as a fundamental step in the formation of memory content (Eccles 1983; Voronin 1983). When ghrelin was administered into the CA1 subfield of the rat hippocampus, a decreased threshold to produce LTP in the dentate gyrus was observed (Carlini et al. 2010b). In addition, ghrelin administration was found to increase hippocampal spine synapse density; ghrelin knockout mice, when compared to their wild-type littermates, had a significantly lower number of dendritic spines in their CA1 hippocampal subfield, an effect that could be almost reversed by treating the knockout mice with exogenous ghrelin (Diano et al. 2006). In the same study, the positive effects of ghrelin and the ghrelin receptor agonist LY444711 on learning and memory were confirmed.

Several studies have demonstrated that ghrelin affects neurogenesis in different brain structures (Zhang 2004; Zhang et al. 2005; Sato et al. 2006), among them the hippocampus. Both in vitro and in vivo administration of ghrelin has been shown to induce proliferation of adult mouse and rat hippocampal progenitor cells (Johansson et al. 2008; Moon et al. 2009; Chung et al. 2013; Li et al. 2013). Since in particular hippocampal neurogenesis has been implicated in learning and memory processes (Zhao et al. 2008), this points to another potential mechanism of ghrelin's effects on memory.

Besides the hippocampus, several other structures induce positive effects on memory processes when receiving direct injections of ghrelin: administration of ghrelin to the amygdala and dorsal raphe nucleus significantly improved memory retention in a step-down avoidance task, even though administration to the hippocampus led to the most sensitive reaction (Carlini et al. 2004). More recent investigations confirmed a role of the amygdala in mediating ghrelin's effects on memory; in a step-through avoidance paradigm ghrelin significantly enhanced memory retention when administered directly to the basolateral amygdala (Goshadrou und Ronaghi 2012).

Mechanisms of Action and Intracellular Signaling Pathways

With ghrelin's role in cognition being gradually defined from a behavioral and neuroanatomical perspective, several biochemical mediators were shown to be involved in the memory enhancing effects of ghrelin. The first transmitter that was found to be an important mediator in ghrelin's impacts on memory was serotonin, whose inhibiting effect on food intake had long been known (Leibowitz und Shor-Posner 1986). Ghrelin significantly and dose-dependently reduced depolarization-induced serotonin release from rat hippocampal synaptosomes (Brunetti et al. 2002). A similar in vitro effect has been seen recently with hippocampal slices (Ghersi et al. 2011). Consistent with these results, selective serotonin reuptake inhibitors could be shown to interfere with the ghrelin-induced enhancement of memory retention. Rats pretreated with fluoxetine did not show a better memory performance after ghrelin administration compared to their untreated mates, neither in an avoidance task nor in an recognition task (Carlini et al. 2007). In order to further explain the networks leading to this interaction between ghrelin and serotonin in memory retention, an influence of serotonin on neuropeptide Y (NPY) levels was proposed (Carlini et al. 2007). NPY is involved both in memory processes and hunger regulation (Flood et al. 1989), essentially mediating ghrelin's orexigenic effects (Kamegai et al. 2001; Shintani et al. 2001). Recent studies indeed confirm an interaction of serotonin and NPY (Crespi 2011; Yada et al. 2012; Bonn et al. 2013).

The effect of ghrelin on LTP seems to be mediated by the biologic messenger nitric oxide (NO); when pretreated with a NO-synthase (NOS) inhibitor, rats did not show better memory retention in an inhibitory avoidance task (Carlini et al. 2010b). For the activation of NOS, calcium (Ca^{2+}) is needed (Bredt and Snyder 1990), and NMDA receptors, regulating the influx of Ca^{2+} into neurons (Crowder et al. 1987), are necessary for the generation of LTP in the hippocampus (Harris et al. 1984). While the effect of ghrelin on voltage-gated Ca^{2+} currents in general is not entirely clear yet (Han et al. 2011; Yamazaki et al. 2004), an interaction between the ghrelin and the NMDA receptor was hypothesized (Carlini et al. 2010b). Recent evidence indeed suggests a close interaction between the two, as certain ghrelin-triggered pathways (see below) seem to result in the enhancement of NMDA receptor function via the phosphorylation of NR1 subunits of this receptor (Cuellar and Isokawa 2011). A synergism seems to be possible as ghrelin acts via a G-protein q (G_q)-coupled pathway regulating intracellular levels of the second messenger inositol trisphosphate (IP_3) (Holst et al. 2003), thus affecting Ca^{2+} availability in the cell (Yamazaki et al. 2012), possibly via IP_3-receptors of the endoplasmic reticulum (Dimitrova et al. 2007; Parys and Smedt 2012). However, antagonism of IP_3-receptors with Xestospongin-C was not selective for inhibiting the ghrelin-induced upregulation of cyclic adenosine monophosphate (cAMP) response element-binding protein (CREB, see below), and thapsigargin, an inhibitor of a Ca^{2+}-pump for the endoplasmic reticulum, did not have any effect on ghrelin increasing intracellular Ca^{2+}-levels (Isokawa 2012). These results indicate that further research is needed to elucidate the role of the ghrelin receptor GHSR in established cellular signal pathways, particularly in its interactions with intracellular Ca^{2+}.

Generally seen as a central biochemical hub for memory formation, the transcription factor CREB has to be switched on via elevated cAMP levels and protein kinase A (PKA) in order to activate a number of genes relevant for memory processes (Gass et al. 1998; Kida 2012). Several signaling molecules influence this process (Morgado-Bernal 2011), and also ghrelin has been shown to be involved in the regulation of CREB activity (Holst et al. 2003). Although the ghrelin receptor GHSR is primarily coupled to a G_q-dependent pathway, ghrelin has been demonstrated to increase CREB activity via the cAMP/PKA-pathway (Cuellar and Isokawa 2011). The enzyme adenylyl cyclase is responsible for the synthesis of cAMP. As the Ca^{2+}-sensitive subtypes AC1 and AC8 of this enzyme have been shown to impact memory and learning via the CREB cascade (Wang und Zhang 2012), it seems likely that ghrelin's impacts on the CREB cascade is mediated via these subtypes.

The ghrelin receptor 1a subtype of the ghrelin receptor has been shown to increase cAMP levels in vitro via augmenting dopamine-induced cAMP accumulation (Jiang 2006). As a neuroanatomical correlate of this effect, the ghrelin receptor 1a and dopamine D1 receptors have been found to be co-expressed in the ventral tegmental area, substantia nigra, and the hippocampus (Jiang 2006). This shows that ghrelin interacts with another transmitter system relevant for learning, in this case indirectly modulating signal transduction rather than directly inducing it. Consistently with this result, it was described that the D1 receptor-antagonist SFK83566 reverses the enhancing effects of ghrelin on an object location memory task (Jacoby and Currie 2011).

Ghrelin and Memory: Linking Belly and Brain

A growing body of evidence suggests that ghrelin, besides its well-described role in hunger regulation, exerts distinct modulatory effects on cognition in general and memory processes in particular. It is part of a multifaceted network of signaling molecules such as leptin, insulin, NPY/AgRP, and many others, interacting to link an organism's energy homeostasis with central processes of learning, memory, and behavior (Bennett et al. 1997; Redrobe et al. 1999; Sarrar et al. 2011; Ghasemi et al. 2013; Warren et al. 2012). From an evolutionary perspective, it makes sense to have such a sumptuous regulatory system involving peripheral and central functions for the act of feeding (Banks 2012). For animals who are forced to survive in the wild—we think again of the squirrel mentioned at the beginning of this chapter—the question of what to eat, when, and where is a highly complex one, as aspects of energy expenditure, cost-benefit analysis, and the presence of natural enemies have to be taken into consideration. Animals are, indeed, able to do so and sometimes show an amazingly cunning behavior when it comes to finding palatable food (Janmaat et al. 2006) and a warm (Balasko und Cabanac 1998) and safe (Pravosudov 2008) place to feed.

Of the peptides mentioned above, ghrelin is the only one synthesized in the stomach (Kojima et al. 1999), its impact on numerous brain regions reaches far beyond the hypothalamic regulation of hunger and satiety: ghrelin takes a role in the regulation of both central circuits involved in food intake and such involved in cognitive functions. For citizens of industrialized societies living in a situation of abundance, the act of eating has become as easy as to open a fridge and to enjoy the permanent availability of a vast variety of food. Hence, the highly developed and sensitive regulatory systems cited above sometimes are led to their limits—with enormous epidemiological implications. Soon after its discovery, ghrelin was seen as a possible target for the treatment of obesity (Dhillo and Bloom 2001). Recent studies focus on its role in diseases such as Parkinson's (Unger et al. 2011) or Alzheimer's (Dos Santos et al. 2013). This shows that understanding both metabolic and neurological conditions such as diabetes and obesity on the one hand as well as dementia and dyskinesia on the other hand needs a comprehensive approach. The elucidation of the mechanisms underlying ghrelin's role in processes of neuroplasticity, neuroprotection, and generally, cognitive processing has therefore important implications for the therapy and prophylaxis of a number of diseases.

References

Albarran-Zeckler RG, Brantley AF, Smith RG (2012) Growth hormone secretagogue receptor (GHSR-1A) knockout mice exhibit improved spatial memory and deficits in contextual memory. Behav Brain Res 232:13–19

Atcha Z, Chen W-S, Ong AB et al (2009) Cognitive enhancing effects of ghrelin receptor agonists. Psychopharmacology 206:415–427

Balasko M, Cabanac M (1998) Behavior of juvenile lizards (Iguana iguana) in a conflict between temperature regulation and palatable food. Brain Behav Evol 52:257–262

Banks WA (2012) Role of the blood-brain barrier in the evolution of feeding and cognition. Ann N Y Acad Sci 1264:13–19

Bayliss JA, Andrews ZB (2013) Ghrelin is neuroprotective in Parkinson's disease: molecular mechanisms of metabolic neuroprotection. Ther Adv Endocrinol Metab 4:25–36

Bennett GW, Ballard TM, Watson CD et al (1997) Eff neuropeptides cogn funct. Exp Gerontol 32:451–469

Bonn M, Schmitt A, Lesch K-P et al (2013) Serotonergic innervation and serotonin receptor expression of NPY-producing neurons in the rat lateral and basolateral amygdaloid nuclei. Brain Struct Funct 218:421–435

Bredt DS, Snyder SH (1990) Isolation of nitric oxide synthetase, a calmodulin-requiring enzyme. Proc Natl Acad Sci USA 87:682–685

Brunetti L, Recinella L, Orlando G et al (2002) Effects of ghrelin and amylin on dopamine, norepinephrine and serotonin release in the hypothalamus. Eur J Pharmacol 454:189–192

Carlini VP, Gaydou RC, Schiöth HB et al (2007) Selective serotonin reuptake inhibitor (fluoxetine) decreases the effects of ghrelin on memory retention and food intake. Regul Pept 140:65–73

Carlini VP, Ghersi M, Schiöth HB et al (2010a) Ghrelin and memory: differential effects on acquisition and retrieval. Peptides 31:1190–1193

Carlini VP, Martini AC, Schiöth HB et al (2008) Decreased memory for novel object recognition in chronically food-restricted mice is reversed by acute ghrelin administration. Neuroscience 153:929–934

Carlini VP, Monzón ME, Varas MM et al (2002) Ghrelin increases anxiety-like behavior and memory retention in rats. Biochem Biophys Res Commun 299:739–743

Carlini VP, Perez MF, Salde E et al (2010b) Ghrelin induced memory facilitation implicates nitric oxide synthase activation and decrease in the threshold to promote LTP in hippocampal dentate gyrus. Physiol Behav 101:117–123

Carlini VP, Varas MM, Cragnolini AB et al (2004) Differential role of the hippocampus, amygdala, and dorsal raphe nucleus in regulating feeding, memory, and anxiety-like behavioral responses to ghrelin. Biochem Biophys Res Commun 313:635–641

Carvajal P, Carlini VP, Schiöth HB et al (2009) Central ghrelin increases anxiety in the Open Field test and impairs retention memory in a passive avoidance task in neonatal chicks. Neurobiol Learn Mem 91:402–407

Chung H, Li E, Kim Y et al (2013) Multiple signaling pathways mediate ghrelin-induced proliferation of hippocampal neural stem cells. J Endocrinol 218:49–59

Crespi F (2011) Influence of Neuropeptide Y and antidepressants upon cerebral monoamines involved in depression: an in vivo electrochemical study. Brain Res 1407:27–37

Crowder JM, Croucher MJ, Bradford HF et al (1987) Excitatory amino acid receptors and depolarization-induced Ca^{2+} influx into hippocampal slices. J Neurochem 48:1917–1924

Cuellar JN, Isokawa M (2011) Ghrelin-induced activation of cAMP signal transduction and its negative regulation by endocannabinoids in the hippocampus. Neuropharmacology 60:842–851

Dhillo WS, Bloom SR (2001) Hypothalamic peptides as drug targets for obesity. Curr Opin Pharmacol 1:651–655

Diano S, Farr SA, Benoit SC et al (2006) Ghrelin controls hippocampal spine synapse density and memory performance. Nat Neurosci 9:381–388

Dimitrova DZ, Mihov DN, Wang R et al (2007) Contractile effect of ghrelin on isolated guinea-pig renal arteries. Vascul Pharmacol 47:31–40

Dos Santos V V, Rodrigues A L S, Lima T C de et al (2013) Ghrelin as a neuroprotective and palliative agent in Alzheimer's and parkinson's disease. Curr Pharm Des 19(38):6773–6790

Dresler M, Kluge M, Genzel L et al (2010) Nocturnal administration of ghrelin does not promote memory consolidation. Pharmacopsychiatry 43:277–278

Eccles JC (1983) Calcium in long-term potentiation as a model for memory. Neuroscience 10:1071–1081

Flood JF, Baker ML, Hernandez EN et al (1989) Modulation of memory processing by neuropeptide Y varies with brain injection site. Brain Res 503:73–82

Gahete MD, Córdoba-Chacón J, Kineman RD et al (2011) Role of ghrelin system in neuroprotection and cognitive functions: implications in Alzheimer's disease. Peptides 32:2225–2228

Gahete MD, Rubio A, Córdoba-Chacón J et al (2010) Expression of the ghrelin and neurotensin systems is altered in the temporal lobe of Alzheimer's disease patients. J Alzheimers Dis 22:819–828

Gass P, Wolfer DP, Balschun D et al (1998) Deficits in memory tasks of mice with CREB mutations depend on gene dosage. Learn Mem 5:274–288

Ghasemi R, Haeri A, Dargahi L et al (2013) Insulin in the Brain: sources, localization and functions. Mol Neurobiol 47:145–171

Ghersi MS, Casas SM, Escudero C et al (2011) Ghrelin inhibited serotonin release from hippocampal slices. Peptides 32:2367–2371

Goshadrou F, Kermani M, Ronaghi A et al (2013) The effect of ghrelin on MK-801 induced memory impairment in rats. Peptides 44:60–65

Goshadrou F, Ronaghi A (2012) Attenuating the effect of Ghrelin on memory storage via bilateral reversible inactivation of the basolateral amygdale. Behav Brain Res 232:391–394

Han X, Zhu Y, Zhao Y et al (2011) Ghrelin reduces voltage-gated calcium currents in GH3 cells via cyclic GMP pathways. Endocrine 40:228–236

Harris EW, Ganong AH, Cotman CW (1984) Long-term potentiation in the hippocampus involves activation of N-methyl-D-aspartate receptors. Brain Res 323:132–137

Holst B, Cygankiewicz A, Jensen TH et al (2003) High constitutive signaling of the ghrelin receptor–identification of a potent inverse agonist. Mol Endocrinol 17:2201–2210

Isokawa M (2012) Cellular signal mechanisms of reward-related plasticity in the hippocampus. Neural Plasticity 2012:1–18

Jacoby SM, Currie PJ (2011) SKF 83566 attenuates the effects of ghrelin on performance in the object location memory task. Neurosci Lett 504:316–320

Janmaat KRL, Byrne RW, Zuberbühler K (2006) Primates take weather into account when searching for fruits. Curr Biol 16:1232–1237

Jiang H (2006) Ghrelin amplifies dopamine signaling by cross talk involving formation of growth hormone secretagogue receptor/dopamine receptor subtype 1 heterodimers. Mol Endocrinol 20:1772–1785

Johansson I, Destefanis S, Aberg ND et al (2008) Proliferative and protective effects of growth hormone secretagogues on adult rat hippocampal progenitor cells. Endocrinology 149:2191–2199

Kamegai J, Tamura H, Shimizu T et al (2001) Chronic central infusion of ghrelin increases hypothalamic neuropeptide Y and Agouti-related protein mRNA levels and body weight in rats. Diabetes 50:2438–2443

Kenny R, Cai G, Bayliss J A et al (2013) Endogenous ghrelin's role in hippocampal neuroprotection after global cerebral ischemia: does endogenous ghrelin protect against global stroke? Am J Physiol Regul Integr Comp Physiol 304(11):R980–990 (1 Jun 2013)

Kida S (2012) A functional role for CREB as a positive regulator of memory formation and LTP. Exp Neurobiol 21:136

Kojima M, Hosoda H, Date Y et al (1999) Ghrelin is a growth-hormone-releasing acylated peptide from stomach. Nature 402:656–660

Laviano A, Molfino A, Rianda S et al (2012) The growth hormone secretagogue receptor (GHSR). Curr Pharm Des 18:4749–4754

Leibowitz SF, Shor-Posner G (1986) Brain serotonin eat behav. Appetite 7:1–14

Li E, Chung H, Kim Y et al (2013) Ghrelin directly stimulates adult hippocampal neurogenesis: implications for learning and memory. Endocr J 60(6):781–789 (ePub 15 Feb 2013)

Moon M, Kim S, Hwang L et al (2009) Ghrelin regulates hippocampal neurogenesis in adult mice. Endocr J 56:525–531

Morgado-Bernal I (2011) Learning and memory consolidation: linking molecular and behavioral data. Neuroscience 176:12–19

Neves G, Cooke SF, Bliss TVP (2008) Synaptic plasticity, memory and the hippocampus: a neural network approach to causality. Nat Rev Neurosci 9:65–75

Parys JB, de Smedt H (2012) Inositol 1,4,5-trisphosphate and its receptors. Adv Exp Med Biol 740:255–279

Pravosudov VV (2008) Mountain chickadees discriminate between potential cache pilferers and non-pilferers. Proc Biol Sci 275:55–61

Redrobe JP, Dumont Y, St-Pierre JA et al (1999) Multiple receptors for neuropeptide Y in the hippocampus: putative roles in seizures and cognition. Brain Res 848:153–166

Sarrar L, Ehrlich S, Merle JV et al (2011) Cognitive flexibility and Agouti-related protein in adolescent patients with anorexia nervosa. Psychoneuroendocrinology 36:1396–1406

Sato M, Nakahara K, Goto S et al (2006) Effects of ghrelin and des-acyl ghrelin on neurogenesis of the rat fetal spinal cord. Biochem Biophys Res Commun 350:598–603

Shintani M, Ogawa Y, Ebihara K et al (2001) Ghrelin, an endogenous growth hormone secretagogue, is a novel orexigenic peptide that antagonizes leptin action through the activation of hypothalamic neuropeptide Y/Y1 receptor pathway. Diabetes 50:227–232

Spitznagel MB, Benitez A, Updegraff J et al (2010) Serum ghrelin is inversely associated with cognitive function in a sample of non-demented elderly. Psychiatry Clin Neurosci 64:608–611

Theodoropoulou A, Metallinos IC, Psyrogiannis A et al (2012) Ghrelin and leptin secretion in patients with moderate Alzheimer's disease. J Nutr Health Aging 16:472–477

Turner E (1969) Hippocampus and memory. Lancet 2:1123–1126

Unger MM, Möller JC, Mankel K et al (2011) Postprandial ghrelin response is reduced in patients with Parkinson's disease and idiopathic REM sleep behaviour disorder: a peripheral biomarker for early Parkinson's disease? J Neurol 258:982–990

Voronin LL (1983) Long-term potentiation in the hippocampus. Neuroscience 10:1051–1069

Wang H, Zhang M (2012) The role of Ca^{2+}-stimulated adenylyl cyclases in bidirectional synaptic plasticity and brain function. Rev Neurosci 23:67–78

Warren MW, Hynan LS, Weiner MF (2012) Leptin and cognition. Dement Geriatr Cogn Disord 33:410–415

Yada T, Kohno D, Maejima Y et al (2012) Neurohormones, rikkunshito and hypothalamic neurons interactively control appetite and anorexia. Curr Pharm Des 18:4854–4864

Yamazaki M, Aizawa S, Tanaka T et al (2012) Ghrelin increases intracellular Ca^{2+} concentration in the various hormone-producing cell types of the rat pituitary gland. Neurosci Lett 526:29–32

Yamazaki M, Kobayashi H, Tanaka T et al (2004) Ghrelin-induced growth hormone release from isolated rat anterior pituitary cells depends on intracellullar and extracellular Ca^{2+} sources. J Neuroendocrinol 16:825–831

Zhang W (2004) Ghrelin stimulates neurogenesis in the dorsal motor nucleus of the vagus. J Physiol 11:2280–2288

Zhang W, Hu Y, Lin TR et al (2005) Stimulation of neurogenesis in rat nucleus of the solitary tract by ghrelin. Peptides 26:2280–2288

Zhao C, Deng W, Gage FH (2008) Mechanisms and functional implications of adult neurogenesis. Cell 132:645–660

Ghrelin Receptors and Epilepsy

Jeanelle Portelli, Ann Massie, Jessica Coppens and Ilse Smolders

Abstract Epilepsy is a neurological disorder that affects more than 50 million people worldwide. One-third of all epilepsy patients do not respond to the anti-epileptic medications that are currently available. As such, there is a great need for new, more effective drugs for the treatment of epilepsy. Drugs that target neuropeptide systems in the brain show great promise for preventing seizures and epilepsy. Little is currently known about the ghrelin receptor and its role in epilepsy. In this chapter, we discuss whether the ghrelin system is a promising target to stop seizures or prevent the development of epilepsy. This is done by looking at what is currently known, as well as what physiological functions of the ghrelin receptor may be beneficial in epilepsy. The final part of this chapter highlights a number of factors that need to be investigated to understand better the function of the ghrelin receptor in epileptic states. These suggestions may indirectly give an insight to researchers studying ghrelin in other fields of research.

Keywords Epilepsy · Seizures · Hippocampus · Heterodimerization · Inflammation · Neuroprotection · Constitutive receptor activity · Ghrelin receptor

Epilepsy is the most common serious neurological disorder worldwide. It can affect anyone, irrelevant of age, racial, geographic or socio-economic boundaries. Around 50 million persons worldwide have active epilepsy with recurrent seizures

J. Portelli · A. Massie · J. Coppens · I. Smolders (✉)
Department of Pharmaceutical Chemistry, Center for Neurosciences, Drug Analysis and Drug Information, Vrije Universiteit Brussel, Laarbeeklaan 103, 1090 Brussels, Belgium
e-mail: ilse.smolders@vub.ac.be

J. Portelli
Laboratory for Clinical and Experimental Neurophysiology, Neurobiology and Neuropsychology, Department of Neurology, Institute for Neuroscience, Ghent University Hospital, De Pintelaan 185, 1K12A, 9000 Gent, Belgium

J. Portelli and I. Smolders (eds.), *Central Functions of the Ghrelin Receptor*,   177
The Receptors 25, DOI: 10.1007/978-1-4939-0823-3_11,
© Springer Science+Business Media New York 2014

(WHO 2009), 30 % of which do not respond to the available medical treatments. This is worrying since epilepsy increases a person's risk of premature death by two-to-three times when compared to the general population, consequently resulting in an urgent need to identify new ways to treat or control this condition.

Epilepsy is defined as a state of recurrent, spontaneous seizures, whereas seizures are generally defined as a period of abnormal, synchronous excitation of a neuronal population. The latter usually lasts for seconds or minutes, however, in the case of status epilepticus (SE) they can be prolonged and continuous (Scharfman 2007). There is a whole field on the study of epileptogenesis, which refers to a dynamic process that with time alters neuronal excitability, establishes critical interconnections and possibly requires complex structural changes in the brain prior to the occurrence of the first spontaneous seizure (Pitkanen and Lukasiuk 2011). In other words, epileptogenesis is the process of converting a normal brain into an epileptic one that supports spontaneous seizures.

Despite all the progress in epilepsy research, researchers are still a long way from understanding the mechanisms underlying seizure generation and epileptogenesis. There has been a lot of improvement in the treatment of patients with epilepsy when compared to the late 1960s, however, a substantial percentage of pharmacoresistant patients still resides (Fattore and Perucca 2011). The majority of antiepileptic drugs (AEDs) in the market act via voltage-dependent ion channels or the GABAergic system (Bialer and White 2010). The ultimate goal of sustained seizure freedom is, however, rarely achieved with the current AEDs on the market, thus the search for better agents is continuously ongoing (Prunetti and Perucca 2011). Major importance is therefore, directed towards the identification of compounds that act in new ways and on novel molecular targets (Rogawski 2006b). Animal models are still essential in the discovery of new AEDs that do not fall under the 'me-too' category and that offer better tolerability, less drug interactions and improved pharmacokinetic interactions (Rogawski 2006a). Ghrelin is a pleiotropic peptide that has gained a lot of attention as a brain–gut hormone. Ghrelin is best known for its role in feeding behaviour and metabolism as explained in Chapters Central Ghrelin Receptors and Food Intake and Ghrelin Receptors a Novel Target for Obesity of this book. Ghrelin also affects other physiological processes in the body, such as pituitary hormone secretion, the cardiovascular system, the autonomic nervous system, the immune system, the musculoskeletal system, memory and sleep regulation, amongst others (Angelidis et al. 2010). The ghrelin system also plays a role in a number of CNS disorders, including epilepsy (Portelli et al. 2012a, b). Ghrelin receptor expression and ghrelin binding sites are present in seizure prone regions of the brain, such as hippocampus and cerebral cortex (Cowley et al. 2003; Diano et al. 2006).

The Ghrelin Receptor Axis and Epilepsy: Preclinical and Clinical Data

The results of the clinical studies performed are contradicting with regard to the direction of plasma ghrelin level alterations in epileptic patients. In the majority of studies, a decrease in ghrelin plasma levels was observed in epileptic patients, which is in line with what was observed in animals (Ataie et al. 2011). The major drawback of these data is that too many variables are present amongst these studies, which can ultimately lead to confusion. Some studies do not specify whether acylated ghrelin or total ghrelin (which also includes des-acyl ghrelin) were analysed, making it difficult to assess whether there were any changes in acylated ghrelin. For a more detailed analysis, one can refer to our previous review (Portelli et al. 2012a). No studies have been performed to date that analyse whether any changes in ghrelin receptor 1a and ghrelin receptor 1b are present in human epileptic brain subjects when compared to control subjects.

Published rodent studies are simpler to interpret since the majority state that ghrelin has anticonvulsant properties (Aslan et al. 2009; Lee et al. 2010; Obay et al. 2007) and there is more information with regard to the role of the ghrelin receptor 1a in epileptic mechanisms. The first group that linked the ghrelin axis to epilepsy reported that a range of intraperitoneal (ip) injections of ghrelin successfully delayed or prevented the development of pentylenetetrazole (PTZ)-induced epileptic seizures in rats (Obay et al. 2007). Obay et al. also showed that oxidative stress, which is known to increase in epileptic seizures, was decreased when rats were pre-treated with ghrelin prior to PTZ administration (Obay et al. 2008b).

Another group has found that intracerebroventricular (i.c.v.) ghrelin presented a U-shaped dose-effect, i.e. the frequency of penicillin-induced epileptiform activity in rats was significantly decreased only following i.c.v. administration of 1 µg and not at 0.5 or 2 µg ghrelin (Aslan et al. 2009). The authors theorised that this U-shaped response could be due to the ability of the ghrelin receptor 1a to rapidly desensitise, or else due to the existence of high and low affinity ghrelin receptor 1a binding sites on different pathways. This study also reported that the anticonvulsant effect of ghrelin required activation of endothelial-NOS/NO route in the brain (Aslan et al. 2009).

Using the chemoconvulsant systemic pilocarpine model for temporal lobe epilepsy (TLE), a recent ex vivo study showed that ghrelin was also found to possess neuroprotective properties by promoting the phosphoinositide 3-kinase (PI3 K)/ Akt signalling pathway and so reversing the decreased ratio of Bcl-2 to Bax induced by seizures, and inhibiting caspase-3 activation (Xu et al. 2009). The authors were unable to confirm whether the neuroprotective effects of ghrelin were due to its action on ghrelin receptor 1a or else another unknown receptor. Indeed, others have already indicated the possibility that ghrelin's anti-apoptotic effects were independent of ghrelin receptor 1a (Delhanty et al. 2007; Granata et al. 2007). In another study using mice, the effect of ghrelin on kainic acid (KA)-induced seizure activity was successfully blocked using the ghrelin receptor 1a antagonist D-Lys3-GHRP6 (Lee et al. 2010). The same group also noted that ghrelin showed anti-apoptopic and anti-inflammatory effects in KA-induced hippocampal neurodegeneration through ghrelin receptor 1a activation.

Recently we have attempted to unravel ghrelin's anticonvulsant mechanism of action using the in vivo rat model for focal pilocarpine-induced limbic seizures, the mouse pilocarpine tail infusion model, transgenic mice with a ghrelin receptor deletion, electrophysiology in hippocampal slices, EEG recording in freely moving rats, and HEK293 cells expressing the human ghrelin receptor (Portelli et al. 2012b). Ghrelin and the ghrelin-mimetic capromorelin attenuated pilocarpine-induced seizures in rats and mice. Experiments with transgenic mice established that ghrelin requires the ghrelin receptor for its anticonvulsant effect. Interestingly we found that ghrelin receptor$^{-/-}$ mice had a higher seizure threshold than ghrelin receptor$^{+/+}$ mice when administered the muscarinic agonist pilocarpine. This prompted us to look further into pharmacological modulation of the receptor where we discovered that abolishing the constitutive activity of ghrelin receptor by inverse agonism results in the attenuation of seizures and epileptiform activity. We verified in HEK293 cells that ghrelin's potential to rapidly desensitize the ghrelin receptor is followed by internalisation of the receptor and a slower resensitization process. This, together with the fact that different ghrelin fragments possess similar agonistic potencies but different desensitisation characteristics on the ghrelin receptor, led us to elucidate that ghrelin probably attenuated limbic seizures in rodents and epileptiform activity in hippocampal slices due to its desensitising effect on the ghrelin receptor (Portelli et al. 2012b). This in turn constituted a novel mechanism of anticonvulsant action whereby an endogenous agonist reduces the activity of a constitutively active receptor.

On the other hand, one study reports that ghrelin was unable to prevent seizures induced by KA or pilocarpine when administered systemically 10 minute prior to chemoconvulsant (Biagini et al. 2011). These authors also assert that des-acyl ghrelin prevented SE in the majority of pilocarpine-treated rats as well as significantly delayed the onset of SE in KA-treated rats.

Ghrelin's Effect on Hippocampal Synaptic Plasticity and Adult Neurogenesis, Possible Relation to Epilepsy

The hippocampus plays a major role in memory formation and repeated seizures can have long-term effects on memory. It has long been regarded that synaptic plasticity changes, notably long-term potentiation (LTP), is crucial in learning and memory processes (Bliss and Collingridge 1993). Human patients suffering from limbic seizures showed altered hippocampal synaptic plasticity, resulting in LTP impairment (Beck et al. 2000). Thus ideally, AEDs should not only efficiently attenuate seizures but as well prohibit memory impairment in TLE patients. Chapter Ghrelin and Memory details on the beneficial role ghrelin receptors have on memory. This could mean that the ghrelin axis, apart from being neuroprotective, could lead to a decrease in memory impairment resulting from epileptic seizures. Chronic models of epilepsy should be performed to investigate the possible antiepileptogenic properties of the ghrelin receptor 1a together with memory preservation.

Neurogenesis persists throughout adulthood in structures involved in TLE such as the hippocampus, albeit at a slower rate than in early life. Aberrant neurogenesis is another factor strongly correlated with hippocampal epileptic tissue (Parent and Kron 2012). Studies in rodent models of medial TLE (mTLE) show that hippocampal neurogenesis is increased during epileptogenesis in post-status models of mTLE, however, decreased in the chronic epileptic stage (Parent and Kron 2012). Moreover, neurogenesis in these models results in abnormalities such as ectopic dentate granule cells and mossy fibre sprouting. These abnormalities may be maladaptive and contribute to the development of the chronic epileptic state and of certain co-morbidities of epilepsy such as depression and memory impairment (Parent and Kron 2012). Very recently, the ghrelin receptor 1a has been found to directly regulate adult hippocampal neurogenesis (Chung et al. 2013; Li et al. 2013; Moon et al. 2009). Exposure of cultured adult rat hippocampal neural stem cells to ghrelin resulted in increased proliferation and increased ghrelin receptor 1a expression, which was attenuated following administration of the ghrelin receptor 1a antagonist D-Lys3-GHRP-6 (Chung et al. 2013). The same group suggested that the effect of ghrelin receptor 1a on neurogenesis may be due to the involvement of the ERK1/2, PI3 K/Akt, and STAT3 signalling pathways. Additionally, the group of Chung implies that PI3 K/Akt-mediated inactivation of GSK-3β and activation of mTOR/p70S6 K may contribute to the proliferative effect of ghrelin (Chung et al. 2013). It is thus of interest to test whether the ghrelin receptor plays a beneficial role in neurogenesis during the process of epileptogenesis, as well as whether ghrelin administration in chronic epileptic rats may lead to an improvement of cognitive abilities via the generation of new neurons.

The Phenomenon of Ghrelin Receptor Heterodimerization and Possible Implications for Epilepsy

The ghrelin receptor has the ability to heteromer with other neurotransmitter receptors, as explained more in depth in Chapter Homodimerization and Heterodimerization of the Ghrelin Receptor. To date, the ghrelin receptor is known to heterodimerize with dopamine 1 (DAD1) and D2 (DAD2) receptors, as well as melanocortin-3 receptors (MC3R) (Jiang et al. 2006; Kern et al. 2012; Rediger et al. 2011). This property of the ghrelin receptor 1a being able to be coexpressed with a different GPCR is still in its initial stages of being understood, and what is known till now is that the properties of each individual receptor can differ when compared to the properties of such receptors present as heterodimers (Kern et al. 2012; Rediger et al. 2011). The implications of such ghrelin receptor

1a co-expression have not yet been studied in the realm of epilepsy, and chances are that these receptor combinations are of importance in such neurological disorders. Not much is known with respect to the role of MC3Rs in epilepsy. It is well accepted that classical neurotransmitter systems can enhance or decrease the threshold for seizure susceptibility. With regards to the dopamine system, DAD1-like and DAD2-like receptors exert opposing actions on intracellular signalling molecules, for instance while DAD1 receptor stimulation activates adenylyl cyclase activity, D2 receptor activation inhibits it (Kebabian and Greengard 1971; Trantham-Davidson et al. 2004). It is generally accepted that activation of DAD1-like receptor family exhibit proconvulsant activity (Gangarossa et al. 2011; O'Sullivan et al. 2008; Starr and Starr 1993) whereas those of the DAD2-like receptor family are anticonvulsant in nature (Clinckers et al. 2004; Starr 1996). Ghrelin receptor 1a is coexpressed with DAD1 receptors in the ventral tegmental area and hippocampus, amongst others, and ghrelin was shown to have the capacity of amplifying hippocampal DAD1 receptor-mediated signalling (Jiang et al. 2006) and extracellular concentrations of accumbal dopamine (Jerlhag et al. 2006). Ghrelin receptor 1a/DAD2 receptor co-expression was found to be present in the hypothalamus, hippocampus and striatum (Kern et al. 2012). One cannot easily hypothesise how such heteromers play a role on epileptic mechanisms. The first step is to investigate whether any changes in expression of such heteromers are present in rodent and human epileptic brains. If such changes in expression are present, this opens a new window in the investigation of heteromers as drug targets in epilepsy.

Inflammation, Neuroprotection and Blood–Brain Barrier Impairment in Epilepsy, Possible Beneficial Roles for Ghrelin Receptor-Mediated Actions

Brain inflammation is thought to play a crucial role in pharmacoresistant epilepsies of different etiologies (Vezzani et al. 2013). Higher levels of inflammatory mediators were detected in both brains of patients suffering from pharmacoresistant epilepsy as well as in animal models of epilepsy (Boer et al. 2008; Choi and Koh 2008; Ravizza et al. 2006, 2008; Vezzani et al. 2011). It is becoming more apparent that the inflammatory cytokines, such as TGF-β and IL-1β, play a detrimental role in the progression of epilepsy; however, their specific roles in epileptogenesis are still under evaluation (Vezzani et al. 2013). A link has been determined between inflammation and disruption of the blood–brain barrier (BBB). A healthy and intact BBB is essential for maintaining an optimal brain environment essential for physiological neuronal function. BBB dysfunction or damage can be a result or can lead to central nervous system diseases and disorders (Marchi et al. 2012; Zlokovic 2008). Seizures were shown to compromise BBB permeability, which could perpetuate or be perpetuated by brain inflammation (Librizzi et al. 2012; van Vliet et al. 2007; Vezzani et al. 2013).

Ghrelin has been shown to act as a potent anti-inflammatory mediator in vivo and in vitro (Baatar et al. 2011; Cheyuo et al. 2011; Stevanovic et al. 2011). In epilepsy settings, ghrelin significantly reduced the accumulation of reactive microglia and astrocytes in the hippocampus following KA-induced excitotoxic injury (Lee et al. 2010). In relation to this, KA-induced increases of Mac-1 (a specific marker for microglial activation) and GFAP (a marker protein for astrogliosis) in the CA1 and CA3 of the hippocampus were potently suppressed by ghrelin. Lee and colleagues also showed that ghrelin inhibited KA-induced increases of TNF-α, IL-1β, COX-2 immunoreactivities as well as Mmp3 expression in the hippocampus (Lee et al. 2010). The mTOR signalling cascade, which has recently been attributed to modulate the process of epileptogenesis (Russo et al. 2012, 2013; Vliet et al. 2012), has also been linked to inflammation (Dello Russo et al. 2009). Ghrelin has been found to require the mTOR pathway for its hypothalamic orexigenic action (Martins et al. 2012). It would be interesting to determine whether ghrelin acts via the mTOR pathway with regard to inflammation. With regard to the BBB, ghrelin was found to prevent BBB disruption following traumatic brain injury (Lopez et al. 2012, 2011).

Ghrelin receptor 1a activation has been shown to exert neuroprotective effects both peripherally and centrally (Ferrini et al. 2009), and as pointed out previously, it has been implicated that ghrelin is capable of stimulating the ERK1/2 and PI3 K/Akt pathways (Chung et al. 2008). Indeed, ghrelin was found to significantly attenuate pilocarpine-induced neuronal loss in hippocampal CA1 and CA3 regions (Xu et al. 2009). The same study also reported that ghrelin upregulated the seizure-induced decreased levels of phospho-PI3 K p85 and phosphor-Akt in the hippocampus, and reversed the decreased Bcl-2 level and the increased Bax level at 24 h after hippocampal pilocarpine treatment. Pilocarpine-induced caspase-3 activation was also inhibited by ghrelin (Xu et al. 2009). These neuroprotective effects were also seen in hippocampal KA-induced seizures. In mice, ghrelin pretreatment significantly reduced hippocampal neuronal cell death, TUNEL-positive cells as well as caspase-3 expression (Lee et al. 2010). Recently, acylated ghrelin was found to inhibit hippocampal neuronal apoptosis in rats undergoing pilocarpine-induced seizures (Zhang et al. 2013). The ghrelin receptor 1a antagonist D-Lys3-GHRH-6 abolished the protective effects of ghrelin, and the authors hypothesise that ghrelin's protective effects may be due to activation of the PI3 K/Akt pathway.

What Role Does the Ghrelin Receptor Play in Epilepsy?

There are high hopes on the role of the ghrelin axis in epilepsy, thanks to its numerous beneficial physiological properties. We are still scratching the surface with regard to the role the ghrelin receptor plays in epileptic states, and much more needs to be discovered.

The Need for Good Antibodies to Quantify Ghrelin Receptor Expression in Human and Rodent Epileptic Tissue

Till date the role of ghrelin receptor 1a in epilepsy has not been appropriately investigated. Ghrelin receptor 1a mRNA and protein levels showed no significant changes at 24 h after pilocarpine-induced seizures in rodents when compared to the control group (Xu et al. 2009). This does not signify that the ghrelin receptor 1a plays no role in epileptic episodes. Changes in ghrelin receptor 1a mRNA and protein levels may take place on a longer time-period, and thus it is important that these parameters are studied in human epileptic brains and in different rodent models. A major stumbling block that slows these studies down is the lack of reliable, specific antibodies. We recently tried to optimise a protocol for detecting (immunohistochemistry) and quantifying (Western blotting) ghrelin receptor 1a protein expression in epileptic brain tissue. We therefore started by investigating the specificity of three independent, commercially available antibodies by comparing labelling in ghrelin receptor 1a knockout and wildtype tissue/samples. However, in our hands, none of the antibodies labelled specifically for the ghrelin receptor 1a observed the same immunoreactive signal in ghrelin receptor 1a knockout tissue/ samples compared to wildtype tissue/samples (unpublished observations). An elaborate study on the different antibodies and protocols for detecting ghrelin receptor 1a, including correct negative controls such as knockout tissue, would be invaluable. Moreover, there might be a need to develop and characterise new antibodies if none of the currently available antibodies seems to be specific.

What Role Does the Constitutive Activity of the Ghrelin Receptor 1a Play in Epilepsy?

As detailed in Chapter Constitutive Activity of the Ghrelin Receptor, Holst and colleagues discovered that the ghrelin receptor 1a has a high constitutive activity, in that it signals with about 50 % of maximal activity in the absence of its peptide ligands (Holst et al. 2003). It is of great importance to understand well this specific characteristic of the ghrelin receptor 1a when it comes to epilepsy. Due to this remarkable phenomenon, one perhaps should be looking at inverse agonists instead of neutral antagonists to block the effect of ghrelin receptor 1a activation. We have in fact found that seizures are attenuated by the inverse ghrelin receptor 1a agonist [D-Arg^1, D-Phe^5, D-Trp^{7,9},Leu^{11}] Substance P as well as A778193 in the pilocarpine model for limbic seizures (Portelli et al. 2012b). Therefore, one should determine whether the known ghrelin receptor 1a antagonists are indeed neutral antagonists or else have inverse agonist properties. The use of mice genetically knocked out of the ghrelin receptor 1a is of essence since it gives a very clear picture of whether the lack of ghrelin receptor 1a incites or inhibits seizures.

Investigating the Ghrelin Receptor Axis in Different Models of Epileptogenesis and Chronic Epilepsy

In order to attempt decreasing the percentage of pharmacoresistant epilepsy patients, the neuropharmacologist needs not only to try to find AEDs with new mechanisms of action but also to keep in mind what information is currently available on the pathophysiology of epilepsy. It is clear that during the complicated process of epileptogenesis, several different mechanisms are taking place; thus one should ideally identify new compounds that are capable of targeting different pathways simultaneously. The focus of epilepsy researchers is to identify compounds that are not only capable of attenuating seizures (anticonvulsant), but are also antiepileptogenic (can prevent epilepsy) or disease-modifying (halting its progression). As previously mentioned, our understanding of the role of the ghrelin axis in the pathogenesis of epilepsy is incomplete. Targeting the ghrelin receptor 1a has shown to attenuate acute seizures in different models (Lee et al. 2010; Obay et al. 2008a, 2007; Portelli et al. 2012a, b; Xu et al. 2009), and from what is already known with regard to this system's properties in view of inflammatory cascades and BBB protection, it is promising that the ghrelin axis could play a role in epileptogenesis.

Thus it is clear that the next step should focus on identifying whether the ghrelin receptor 1a is involved in the epileptogenesis process or else can be targeted in chronic epilepsy.

Does Heterodimerization of the Ghrelin Receptor Participate in Any Way in Epilepsy?

Another factor that has not been studied yet is the role of the ghrelin receptor 1b in epilepsy. It has been believed for years that ghrelin receptor 1b is inactive, however, this notion was questionable since this isoform is widely spread in different tissues of the body. Nowadays, it is thought to play a significant role in modulating ghrelin receptor 1b and other GPCRs through GPCR homo-and/or heterodimerization, and it is thought to be a negative regulator of ghrelin receptor 1a (Chu et al. 2007; Leung et al. 2007). Indeed, recently Mary et al. discovered that ghrelin receptor 1a restricts the conformational landscape of the full-length ghrelin receptor 1a, rendering the latter receptor inactive (Mary et al. 2013). One should assess whether the expression of this receptor is affected in human epileptic hippocampal brain tissues or not. Apart from this, more prominence should be given to the heterodimerization of the ghrelin receptor 1a to receptors of other families, and assess their implication in both epileptogenesis and fully developed epilepsy.

Concluding Remarks

The ghrelin receptor 1a is a fascinating receptor with numerous physiological functions, which has without a doubt stirred a lot of excitement in the scientific community. We feel that ghrelin ligands have a great clinical potential in the field of epilepsy, however, more needs to be done to fully understand the position of this receptor in epilepsy mechanisms.

References

Angelidis G, Valotassiou V, Georgoulias P (2010) Current and potential roles of ghrelin in clinical practice. J Endocrinol Invest 33:823–838

Aslan A, Yildirim M, Ayyildiz M, Guven A, Agar E (2009) The role of nitric oxide in the inhibitory effect of ghrelin against penicillin-induced epileptiform activity in rat. Neuropeptides 43:295–302

Ataie Z, Golzar MG, Babri S, Ebrahimi H, Mohaddes G (2011) Does ghrelin level change after epileptic seizure in rats? Seizure 20:347–349

Baatar D, Patel K, Taub DD (2011) The effects of ghrelin on inflammation and the immune system. Mol Cell Endocrinol 340:44–58

Beck H, Goussakov IV, Lie A, Helmstaedter C, Elger CE (2000) Synaptic plasticity in the human dentate gyrus. J Neurosci 20:7080–7086

Biagini G, Torsello A, Marinelli C, Gualtieri F, Vezzali R, Coco S, Bresciani E, Locatelli V (2011) Beneficial effects of desacyl-ghrelin, hexarelin and EP-80317 in models of status epilepticus. Eur J Pharmacol 670:130–136

Bialer M, White HS (2010) Key factors in the discovery and development of new antiepileptic drugs. Nat Rev Drug Discov 9:68–82

Bliss TV, Collingridge GL (1993) A synaptic model of memory: long-term potentiation in the hippocampus. Nature 361:31–39

Boer K, Jansen F, Nellist M, Redeker S, van den Ouweland AM, Spliet WG, van Nieuwenhuizen O, Troost D, Crino PB, Aronica E (2008) Inflammatory processes in cortical tubers and subependymal giant cell tumors of tuberous sclerosis complex. Epilepsy Res 78:7–21

Cheyuo C, Wu R, Zhou M, Jacob A, Coppa G, Wang P (2011) Ghrelin suppresses inflammation and neuronal nitric oxide synthase in focal cerebral ischemia via the vagus nerve. Shock 35:258–265

Choi J, Koh S (2008) Role of brain inflammation in epileptogenesis. Yonsei Med J 49:1–18

Chu KM, Chow KB, Leung PK, Lau PN, Chan CB, Cheng CH, Wise H (2007) Over-expression of the truncated ghrelin receptor polypeptide attenuates the constitutive activation of phosphatidylinositol-specific phospholipase C by ghrelin receptors but has no effect on ghrelin-stimulated extracellular signal-regulated kinase 1/2 activity. Int J Biochem Cell Biol 39:752–764

Chung H, Seo S, Moon M, Park S (2008) Phosphatidylinositol-3-kinase/Akt/glycogen synthase kinase-3 beta and ERK1/2 pathways mediate protective effects of acylated and unacylated ghrelin against oxygen-glucose deprivation-induced apoptosis in primary rat cortical neuronal cells. J Endocrinol 198:511–521

Chung H, Li E, Kim Y, Kim S, Park S (2013) Multiple signaling pathways mediate ghrelin-induced proliferation of hippocampal neural stem cells. J Endocrinol 218(1):49–59

Clinckers R, Smolders I, Meurs A, Ebinger G, Michotte Y (2004) Anticonvulsant action of hippocampal dopamine and serotonin is independently mediated by D and 5-HT receptors. J Neurochem 89:834–843

Cowley MA, Smith RG, Diano S, Tschop M, Pronchuk N, Grove KL, Strasburger CJ, Bidlingmaier M, Esterman M, Heiman ML, Garcia-Segura LM, Nillni EA, Mendez P, Low MJ, Sotonyi P, Friedman JM, Liu H, Pinto S, Colmers WF, Cone RD, Horvath TL (2003) The distribution and mechanism of action of ghrelin in the CNS demonstrates a novel hypothalamic circuit regulating energy homeostasis. Neuron 37:649–661

Delhanty PJ, van Koetsveld PM, Gauna C, van de Zande B, Vitale G, Hofland LJ, van der Lely AJ (2007) Ghrelin and its unacylated isoform stimulate the growth of adrenocortical tumor cells via an anti-apoptotic pathway. Am J Physiol Endocrinol Metab 293:E302–E309

Dello Russo C, Lisi L, Tringali G, Navarra P (2009) Involvement of mTOR kinase in cytokine-dependent microglial activation and cell proliferation. Biochem Pharmacol 78:1242–1251

Diano S, Farr SA, Benoit SC, McNay EC, da Silva I, Horvath B, Gaskin FS, Nonaka N, Jaeger LB, Banks WA, Morley JE, Pinto S, Sherwin RS, Xu L, Yamada KA, Sleeman MW, Tschop MH, Horvath TL (2006) Ghrelin controls hippocampal spine synapse density and memory performance. Nat Neurosci 9:381–388

Fattore C, Perucca E (2011) Novel medications for epilepsy. Drugs 71:2151–2178

Ferrini F, Salio C, Lossi L, Merighi A (2009) Ghrelin in central neurons. Curr Neuropharmacol 7:37–49

Gangarossa G, Di Benedetto M, O'Sullivan GJ, Dunleavy M, Alcacer C, Bonito-Oliva A, Henshall DC, Waddington JL, Valjent E, Fisone G (2011) Convulsant doses of a dopamine D1 receptor agonist result in Erk-dependent increases in Zif268 and Arc/Arg3.1 expression in mouse dentate gyrus. PLoS ONE 6:e19415

Granata R, Settanni F, Biancone L, Trovato L, Nano R, Bertuzzi F, Destefanis S, Annunziata M, Martinetti M, Catapano F, Ghe C, Isgaard J, Papotti M, Ghigo E, Muccioli G (2007) Acylated and unacylated ghrelin promote proliferation and inhibit apoptosis of pancreatic beta-cells and human islets: involvement of $3',5'$-cyclic adenosine monophosphate/protein kinase A, extracellular signal-regulated kinase 1/2, and phosphatidyl inositol 3-Kinase/Akt signaling. Endocrinology 148:512–529

Holst B, Cygankiewicz A, Jensen TH, Ankersen M, Schwartz TW (2003) High constitutive signaling of the ghrelin receptor–identification of a potent inverse agonist. Mol Endocrinol 17:2201–2210

Jerlhag E, Egecioglu E, Dickson SL, Andersson M, Svensson L, Engel JA (2006) Ghrelin stimulates locomotor activity and accumbal dopamine-overflow via central cholinergic systems in mice: implications for its involvement in brain reward. Addict Biol 11:45–54

Jiang H, Betancourt L, Smith RG (2006) Ghrelin amplifies dopamine signaling by cross talk involving formation of growth hormone secretagogue receptor/dopamine receptor subtype 1 heterodimers. Mol Endocrinol 20:1772–1785

Kebabian JW, Greengard P (1971) Dopamine-sensitive adenyl cyclase: possible role in synaptic transmission. Science 174:1346–1349

Kern A, Albarran-Zeckler R, Walsh HE, Smith RG (2012) Apo-ghrelin receptor forms heteromers with DRD2 in hypothalamic neurons and is essential for anorexigenic effects of DRD2 agonism. Neuron 73:317–332

Lee J, Lim E, Kim Y, Li E, Park S (2010) Ghrelin attenuates kainic acid-induced neuronal cell death in the mouse hippocampus. J Endocrinol 205:263–270

Leung PK, Chow KB, Lau PN, Chu KM, Chan CB, Cheng CH, Wise H (2007) The truncated ghrelin receptor polypeptide (GHS-R1b) acts as a dominant-negative mutant of the ghrelin receptor. Cell Signal 19:1011–1022

Li E, Chung H, Kim Y, Kim DH, Ryu JH, Sato T, Kojima M, Park S (2013) Ghrelin directly stimulates adult hippocampal neurogenesis: implications for learning and memory. Endocr J 60(6):781–789

Librizzi L, Noe F, Vezzani A, de Curtis M, Ravizza T (2012) Seizure-induced brain-borne inflammation sustains seizure recurrence and blood-brain barrier damage. Ann Neurol 72:82–90

Lopez NE, Krzyzaniak MJ, Blow C, Putnam J, Ortiz-Pomales Y, Hageny AM, Eliceiri B, Coimbra R, Bansal V (2011) Ghrelin prevents disruption of the Blood-Brain Barrier after traumatic Brain injury. J Neurotrauma

Lopez NE, Gaston L, Lopez KR, Coimbra RC, Hageny A, Putnam J, Eliceiri B, Coimbra R, Bansal V (2012) Early ghrelin treatment attenuates disruption of the blood brain barrier and apoptosis after traumatic brain injury through a UCP-2 mechanism. Brain Res 1489:140–148

Marchi N, Granata T, Ghosh C, Janigro D (2012) Blood-brain barrier dysfunction and epilepsy: pathophysiologic role and therapeutic approaches. Epilepsia 53:1877–1886

Martins L, Fernandez-Mallo D, Novelle MG, Vazquez MJ, Tena-Sempere M, Nogueiras R, Lopez M, Dieguez C (2012) Hypothalamic mTOR signaling mediates the orexigenic action of ghrelin. PLoS ONE 7:e46923

Mary S, Fehrentz JA, Damian M, Gaibelet G, Orcel H, Verdie P, Mouillac B, Martinez J, Marie J, Baneres JL (2013) Heterodimerization with Its splice variant blocks the ghrelin receptor 1a in a non-signaling conformation: a study with a purified heterodimer assembled into lipid discs. J Biol Chem 288:24656–24665

Moon M, Kim S, Hwang L, Park S (2009) Ghrelin regulates hippocampal neurogenesis in adult mice. Endocr J 56:525–531

Obay BD, Tasdemir E, Tumer C, Bilgin HM, Sermet A (2007) Antiepileptic effects of ghrelin on pentylenetetrazole-induced seizures in rats. Peptides 28:1214–1219

Obay BD, Tasdemir E, Tumer C, Bilgin H, Atmaca M (2008a) Dose dependent effects of ghrelin on pentylenetetrazole-induced oxidative stress in a rat seizure model. Peptides 29:448–455

Obay BD, Tasdemir E, Tumer C, Bilgin HM, Atmaca M (2008b) Dose dependent effects of ghrelin on pentylenetetrazole-induced oxidative stress in a rat seizure model. Peptides 29:448–455

O'Sullivan GJ, Dunleavy M, Hakansson K, Clementi M, Kinsella A, Croke DT, Drago J, Fienberg AA, Greengard P, Sibley DR, Fisone G, Henshall DC, Waddington JL (2008) Dopamine D1 versus D5 receptor-dependent induction of seizures in relation to DARPP-32, ERK1/2 and GluR1-AMPA signalling. Neuropharmacology 54:1051–1061

Parent JM, Kron MM (2012) Neurogenesis and epilepsy. In: Noebels JL, Avoli M, Rogawski MA, Olsen RW, Delgado-Escueta AV (eds) Jasper's basic mechanisms of the epilepsies [Internet], 4th edn. National Center for Biotechnology Information, Bethesda

Pitkanen A, Lukasiuk K (2011) Mechanisms of epileptogenesis and potential treatment targets. Lancet Neurol 10:173–186

Portelli J, Michotte Y, Smolders I (2012a) Ghrelin: an emerging new anticonvulsant neuropeptide. Epilepsia 53:585–595

Portelli J, Thielemans L, Ver Donck L, Loyens E, Coppens J, Aourz N, Aerssens J, Vermoesen K, Clinckers R, Schallier A, Michotte Y, Moechars D, Collingridge GL, Bortolotto ZA, Smolders I (2012b) Inactivation of the constitutively active ghrelin receptor attenuates limbic seizure activity in rodents. Neurotherapeutics 9:658–672

Prunetti P, Perucca E (2011) New and forthcoming anti-epileptic drugs. Curr Opin Neurol 24:159–164

Ravizza T, Boer K, Redeker S, Spliet WG, van Rijen PC, Troost D, Vezzani A, Aronica E (2006) The IL-1beta system in epilepsy-associated malformations of cortical development. Neurobiol Dis 24:128–143

Ravizza T, Gagliardi B, Noe F, Boer K, Aronica E, Vezzani A (2008) Innate and adaptive immunity during epileptogenesis and spontaneous seizures: evidence from experimental models and human temporal lobe epilepsy. Neurobiol Dis 29:142–160

Rediger A, Piechowski CL, Yi CX, Tarnow P, Strotmann R, Gruters A, Krude H, Schoneberg T, Tschop MH, Kleinau G, Biebermann H (2011) Mutually opposite signal modulation by hypothalamic heterodimerization of ghrelin and melanocortin-3 receptors. J Biol Chem 286:39623–39631

Rogawski MA (2006a) Diverse mechanisms of antiepileptic drugs in the development pipeline. Epilepsy Res 69:273–294

Rogawski MA (2006b) Molecular targets versus models for new antiepileptic drug discovery. Epilepsy Res 68:22–28

Russo E, Citraro R, Constanti A, De Sarro G (2012) The mTOR signaling pathway in the brain: focus on epilepsy and epileptogenesis. Mol Neurobiol 46:662–681

Russo E, Citraro R, Donato G, Camastra C, Iuliano R, Cuzzocrea S, Constanti A, De Sarro G (2013) mTOR inhibition modulates epileptogenesis, seizures and depressive behavior in a genetic rat model of absence epilepsy. Neuropharmacology 69:25–36

Scharfman HE (2007) The neurobiology of epilepsy. Curr Neurol Neurosci Rep 7:348–354

Starr MS (1996) The role of dopamine in epilepsy. Synapse 22:159–194

Starr MS, Starr BS (1993) Seizure promotion by D1 agonists does not correlate with other dopaminergic properties. J Neural Transm Park Dis Dement Sect 6:27–34

Stevanovic D, Starcevic V, Vilimanovich U, Nesic D, Vucicevic L, Misirkic M, Janjetovic K, Savic E, Popadic D, Sudar E, Micic D, Sumarac-Dumanovic M, Trajkovic V (2011) Immunomodulatory actions of central ghrelin in diet-induced energy imbalance. Brain Behav Immun 26:150–158

Trantham-Davidson H, Neely LC, Lavin A, Seamans JK (2004) Mechanisms underlying differential D1 versus D2 dopamine receptor regulation of inhibition in prefrontal cortex. J Neurosci 24:10652–10659

Vliet EA van, Costa Araujo S da, Redeker S, Schaik R van, Aronica E, Gorter JA (2007) Blood-brain barrier leakage may lead to progression of temporal lobe epilepsy. Brain 130:521–534

Vezzani A, French J, Bartfai T, Baram TZ (2011) The role of inflammation in epilepsy. Nat Rev Neurol 7:31–40

Vezzani A, Friedman A, Dingledine RJ (2013) The role of inflammation in epileptogenesis. Neuropharmacology 69:16–24

Vliet EA van, Forte G, Holtman L, Burger JC den, Sinjewel A, Vries HE de, Aronica E, Gorter JA. Inhibition of mammalian target of rapamycin reduces epileptogenesis and blood-brain barrier leakage but not microglia activation. Epilepsia, 2012; 53: 1254-63

World Health Organization (2009) Epilepsy in the WHO European region. http://www.ibe-epilepsy.org/downloads/EURO%20Report%20160510.pdf

Xu J, Wang S, Lin Y, Cao L, Wang R, Chi Z (2009) Ghrelin protects against cell death of hippocampal neurons in pilocarpine-induced seizures in rats. Neurosci Lett 453:58–61

Zhang R, Yang G, Wang Q, Guo F, Wang H (2013) Acylated ghrelin protects hippocampal neurons in pilocarpine-induced seizures of immature rats by inhibiting cell apoptosis. Mol Biol Rep 40:51–58

Zlokovic BV (2008) The blood-brain barrier in health and chronic neurodegenerative disorders. Neuron 57:178–201

Ghrelin Plays a Role in Various Physiological and Pathophysiological Brain Functions

Sarah J. Spencer

Abstract The ghrelin receptor is now known to play an important role in regulating physiological responses to stress. In particular, ghrelin acting at the growth hormone secretagogue receptor (ghrelin receptor) may promote anxious behaviours under non-stressed conditions, and attenuate anxiety under conditions of stress. Dysregulation of the ghrelin system therefore has significant consequences for stress-related mood disorders such as anxiety and depression; disorders that pose a substantial problem for human health. These effects of the ghrelin system on mood are of particular concern in obese populations, where the likelihood of a mood disorder is higher and the ghrelin system disrupted. Studies in humans are still revealing conflicting roles for ghrelin and the ghrelin receptor in anxiety and depression, but these, and studies in animal models, offer evidence that ghrelin may influence its receptor at extra-hypothalamic brain regions to exert indirect control over central responses to stress and over brain pathways related to anxiety and depression. In this chapter, I discuss the background and potential mechanisms for ghrelin and ghrelin receptor's role in regulating stress and stress-related mood disorders.

Keywords Anxiety · Depression · Obesity · Hypothalamic–pituitary–adrenal axis · Paraventricular nucleus of the hypothalamus · Stress

The Health Implications of Anxiety and Depression

Anxiety and depression are stress-related mood disorders that pose a significant problem for human health. In the developed world, around 28 % of people will suffer from some type of anxiety or depression-related mood disorder in any one

S. J. Spencer (✉)
School of Health Sciences and Health Innovations Research Institute (HIRi),
RMIT University, Melbourne, VIC, 3083 Australia
e-mail: Sarah.Spencer@rmit.edu.au

J. Portelli and I. Smolders (eds.), *Central Functions of the Ghrelin Receptor*,
The Receptors 25, DOI: 10.1007/978-1-4939-0823-3_12,
© Springer Science+Business Media New York 2014

year and 45 % of people will experience one of these in their lifetime (Anxiety Depression Association of America 2013).

Anxiety and depression are interrelated and both are associated with other adverse health effects such as sleep problems and the use of substances of abuse (Schneiderman et al. 2005). Both these disorders are also strongly stress related. Chronic or severe acute stress can precipitate the onset of these mood disorders. For instance, depressive patients are more likely to have had a stressful life event prior to diagnosis than age-matched controls (Schneiderman et al. 2005). Anxiety and depression are thus strongly associated with hyperactivity of the hypothalamic–pituitary–adrenal (HPA) stress axis (Staufenbiel et al. 2012; Lloyd and Nemeroff 2011).

The HPA axis is the endocrine arm of the body's response to stress (Spencer and Tilbrook 2011; Sapolsky et al. 2000; Papadimitriou and Priftis 2009). When an animal (including humans) is stressed, the arginine vasopressin (AVP)- and corticotropin-releasing hormone (CRH)-containing medial parvocellular cells in the paraventricular nucleus of the hypothalamus (PVN) are activated, and this leads to the release of AVP and CRH into the median eminence where they act on corticotropic cells in the anterior pituitary to stimulate the release of adrenocorticotropic hormone (ACTH) into circulation. ACTH acts on the adrenal cortex to stimulate glucocorticoid release (corticosterone in rodents, cortisol in humans), and this hormone mediates many acutely adaptive functions to facilitate coping with the stress. Glucocorticoids stimulate glucose uptake at the skeletal muscle, mobilise glucose and fat stores, enhance synaptic plasticity to improve learning and memory and suppress cytokine production and thus inflammation. Glucocorticoids also feed back onto the brain to prevent further activation of the HPA axis (Spencer and Tilbrook 2011; Sapolsky et al. 2000; Papadimitriou and Priftis 2009). Dysregulation of this axis is a characterising factor of anxiety and depression (Staufenbiel et al. 2012; Lloyd and Nemeroff 2011). Recent evidence now suggests that ghrelin may be a crucial element in regulating the HPA axis under conditions of stress, and abnormalities in the ghrelin system are likely to contribute to the development of stress-related mood disorders (Lutter et al. 2008; Spencer et al. 2012; Raspopow et al. 2010; Patterson et al. 2013).

Ghrelin and Ghrelin Receptors in Stress-Related Mood Disorders in Humans

Obesity, Ghrelin and Stress-Related Mood Disorders

Obesity is one of the leading causes of death and disease in the developed world, with countries like the USA and Australia reporting as many as 60 % of its citizens are now overweight or obese (Australian Health Survey 2011). In addition to myriad other co-morbidities, there is a well-reported association between obesity and psychiatric disorders, including anxiety and depression (Abiles et al. 2010;

Scott et al. 2008; Doyle et al. 2007). Hyperphagia, a rapid increase in body weight and obesity are linked to major depressive disorder (Novick et al. 2005; Simon et al. 2006; Kloiber et al. 2007), and atypical depression in particular is associated with carbohydrate craving, weight gain and anxiety (Juruena and Cleare 2007). For instance, a body mass index of more than 30 can lead to a 25 % increase in the likelihood a person will develop a mood disorder (Simon et al. 2006; Kloiber et al. 2007). Independent studies have shown teenage girls with depression to have a 2.3-fold increase in the risk of obesity in adulthood (Richardson et al. 2003), and US army veterans with post-traumatic stress disorder to have a 20 % greater rate of obesity than the general US population (Vieweg et al. 2006).

Conversely, or perhaps precipitously, stress also influences feeding behaviour. Although many individuals are inclined to eat less in stressful situations, others overeat and are particularly likely to prefer calorically dense highly palatable foods (Oliver and Wardle 1999; Dallman 2009; Torres and Nowson 2007; Block et al. 2009; Serlachius et al. 2007; Gibson 2006). It is therefore clear that many of the neuropeptides crucial in regulating feeding and metabolism are also involved in stress and mood disorders.

Ghrelin Profiles are Altered in Stress-Related Mood Disorders

Ghrelin is the one of the few hormones known to stimulate feeding and its involvement in stress-related disorders such as depression and anxiety, particularly in obesity, may therefore be very important. A polymorphism in the preproghrelin gene has recently been associated with the anxiety disorder, panic (Hansson et al. 2013). Furthermore, circulating ghrelin levels are altered in some (but not all) studies of major depressive disorders. Thus, some cohorts of patients with major depressive disorder have lower circulating ghrelin levels than control patients (Barim et al. 2009), some cohorts have higher circulating ghrelin concentrations (Kurt et al. 2007), and in others there is no correlation (Kluge et al. 2009; Emul et al. 2007; Schanze et al. 2008). The discrepancies between these studies may be partly due to small sample sizes, but also to how the samples were collected for assessment of ghrelin levels as the peptide hydrolyses quickly after collection in the absence of treatment with EDTA–aprotinin (Hosoda and Kangawa 2012). Additionally, the relative ratios of the acyl and desacyl forms of ghrelin may also be important in influencing stress and mood disorders, and these have so far been assessed in very few studies (Barim et al. 2009).

As we have seen in rodents (Spencer et al. 2012), a background of stress may also have influenced ghrelin levels in these patients to obscure the findings. As such, it is interesting that circulating ghrelin in people who do not overeat when under stress (low-emotional eaters) is higher than that of people who do (high-emotional eaters). Ghrelin also declines in low-emotional eaters but not in high-emotional eaters following food intake, which may explain why the latter keep eating (Raspopow et al. 2010).

In addition to an effect of mood disorders on ghrelin, defects in the receptor have also been linked with depression in some studies. For example, a polymorphism in the GHSR gene has been associated with major depressive disorder, but not with panic disorder in a cohort of patients (Nakashima et al. 2008).

Crucially, manipulating ghrelin levels can lead to changes in the manifestation of depression, even in cases where circulating levels were normal prior to treatment. Thus, a tendency was found for ghrelin to have antidepressive effects in one cohort of patients with major depressive disorder, particularly in regard to ameliorating their sleep disturbances (Kluge et al. 2011). Ghrelin itself can be elevated in the blood with antidepressant treatments, again suggesting it may be able to ameliorate symptoms (Pinar et al. 2008). Conversely, an improvement in major depressive disorder has also been associated with a decrease in circulating ghrelin (Kurt et al. 2007; Barim et al. 2009; Schmid et al. 2006). In this regard, recent study has found ghrelin levels are higher in patients with major depressive disorder who do not respond to treatment and lower in those who do, suggesting reduced ghrelin is beneficial in this case (Ishitobi et al. 2012).

Circulating ghrelin levels also change subject to stress. Thus, plasma ghrelin is increased by about 40 % after 10 min in humans given a Trier Social Stress Test (Raspopow et al. 2010; Rouach et al. 2007). It appears from some studies that ghrelin may stimulate or facilitate the stress response rather than the converse as exogenous ghrelin injections increase the stress hormones cortisol and ACTH, (Takaya et al. 2000; Arvat et al. 2001) and both endogenous and exogenous glucocorticoids cause a reduction in plasma ghrelin (Otto et al. 2004).

Ghrelin's exact role in stress-related mood disorders in humans is not yet clear. It appears that reduced ghrelin or a defect in its ability to interact with the receptor can lead to anxiety and depression and these disorders may be improved by exogenous ghrelin treatment. However, this is not necessarily the case with all studies, and in some reduced ghrelin is associated with beneficial outcomes on depressive symptoms (Ishitobi et al. 2012; Kurt et al. 2007; Barim et al. 2009; Schmid et al. 2006). Animal models may therefore offer a clearer understanding of how ghrelin is acting.

Ghrelin and Ghrelin Receptors in Stress, Anxiety and Depression in Animal Models

Ghrelin Profiles are Altered in Response to Stress in Animal Models

As has been seen in humans, ghrelin increases in response to acute and chronic models of depression and anxiety in rodents (Lutter et al. 2008). For instance, plasma ghrelin and ghrelin mRNA in the gut are increased after repeated tail pinch stress and repeated water avoidance stress in rats (Asakawa et al. 2001; Kristenssson et al. 2006; Ochi et al. 2008). Plasma desacyl and acyl ghrelin, preproghrelin mRNA and numbers of ghrelin-containing cells are increased after chronic daily restraint (Zheng et al. 2009), 14-day chronic unpredictable stress (Patterson et al. 2010) and chronic social defeat stress in rats and mice (Berton et al. 2006; Nestler and Hyman 2010; Lutter et al. 2008). Strikingly, chronic stress can lead to persistently elevated plasma ghrelin. Thus, chronic social defeat stress for 10 days led to increased acylated ghrelin in the plasma and this was still elevated when the mice were assessed one month later (Lutter et al. 2008). As further evidence that ghrelin is closely associated with modulating stress and related mood disorders, Wistar-Kyoto rats are more anxious than Sprague Dawley rats; they also have lower plasma ghrelin and less of a stress-induced elevation of ghrelin (Kristenssson et al. 2006, 2007; Florentzson et al. 2009).

Changes in Ghrelin Profiles are Linked to Anxiety-Like and Depression-Like Behaviour in Animal Models

Animal models of chronic stress such as chronic social defeat, chronic unpredictable stress and chronic daily restraint are commonly used to mimic and study depression. They induce many of the hallmarks of major depressive disorder, including social withdrawal, anorexia or hyperphagia, anhedonia, poor coping in the Porsolt's forced swim test and a reduced exploratory drive (Cryan and Slattery 2007). These behaviours can be reversed with antidepressants (Cryan and Slattery 2007). Chronic social defeat stress involves, for the rodent, several consecutive days of forced social interaction with a larger more dominant conspecific. The test mouse is defeated daily by the larger mouse and comes to display social withdrawal and increased immobility in the forced swim test (Cryan and Slattery 2007; Lutter et al. 2008). These depressive-like behaviours, i.e. social isolation and immobility in the forced swim, are both reversed by elevating plasma ghrelin. Thus, 10 days of calorie restriction (to elevate endogenous ghrelin) or a single subcutaneous (sc) injection of ghrelin reduced immobility in the forced swim test, and reduced anxiety in elevated plus maze induced by chronic social defeat stress (Lutter et al. 2008). Neither the calorie restriction nor the sc injection had any effect in mice lacking the ghrelin receptor, indicating a role for acyl ghrelin at the ghrelin receptor in mitigating depressive and anxious behaviour (Lutter et al. 2008). Ghrelin receptor-null mice also showed more depressive behaviour in a social interaction test. That is, they spent more time socially isolated after the chronic social defeat stress than wild-type mice with normal ghrelin signalling (Lutter et al. 2008). These data seem to show elevating ghrelin after stress may be an adaptive mechanism to assist coping and reduce anxiety and depression (Chuang and Zigman 2010).

Lutter and colleagues report the results of calorie restriction and sc ghrelin. However, other studies have shown centrally applied ghrelin has very different effects. Suppression of central ghrelin action by intracerebroventricular (i.c.v.) administration of antisense ghrelin oligonucleotides caused antidepressive and anxiolytic effects in the forced swim and elevated plus maze in the rat, suggesting ghrelin is normally depressive (Kanehisa et al. 2006). Ghrelin given i.c.v. to mice caused anxiety-like behaviour in the elevated plus maze when the mice were tested 10 min after injection (Asakawa et al. 2001). Ghrelin given i.c.v. or directly into the hippocampus, dorsal raphe nucleus or amygdala also caused anxiety-like behaviour in the elevated plus maze, open-field and step-down inhibitory avoidance tests when the mice were tested 5 min after injection (Carlini et al. 2002, 2004). The involvement of the amygdala, at least, in these behaviours may depend on metabolic status and satiety, as a separate study was able to show ghrelin injected directly into the amygdala can reduce anxiety in the elevated plus maze but only if the rats were food restricted (Alvarez-Crespo et al. 2012).

All these studies have reported findings of acute doses of ghrelin, which may suggest ghrelin has a different role depending on if it is acutely versus chronically elevated. However, Hansson and colleagues have recently reported chronic ghrelin delivered i.c.v. clearly induces anxiety-like behaviour in the open-field and elevated plus maze, and depression-like behaviour in the forced swim test (Hansson et al. 2011).

Many suggestions have been put forward to explain the differences between the various findings, including that ghrelin's role in stress, anxiety and depression may be specific to species, strain, dose, timing, route of administration and metabolic status. While some, or all, of these interpretations may be correct, there is also evidence to suggest ghrelin plays a dual role in stress, anxiety and depression. Thus, findings from our group suggest that ghrelin promotes anxiety under non-stressed conditions and attenuates it following stress (Spencer et al. 2012). Mice lacking endogenous ghrelin (ghr-/-) spent more time in the open arms of the elevated plus maze and in the centre of the open-field than wild-type mice before stress. After stress this profile was reversed, with the ghr-/- now showing more anxiety (Spencer et al. 2012). Thus, low levels of ghrelin, such as under non-stressed conditions in normal animals, promote anxiety-like behaviour. High levels of ghrelin, such as under fasted conditions or stress, attenuate anxiety and depression, consistent with ghrelin's role in enhancing food-seeking (Fig. 1). This idea is in accordance with Lutter et al.'s findings that ghrelin defends against the effects of chronic stress (Lutter et al. 2008) and with Hansson et al.'s finding that chronic ghrelin in the absence of stress enhances anxiety (Hansson et al. 2011). It is also supported by some of the studies of acute ghrelin injection where behavioural tests were conducted under basal (unstressed) conditions after ghrelin injection (Carlini et al. 2002, 2004; Asakawa et al. 2001; Kanehisa et al. 2006). The mechanism by which ghrelin regulates mood is yet unclear, however.

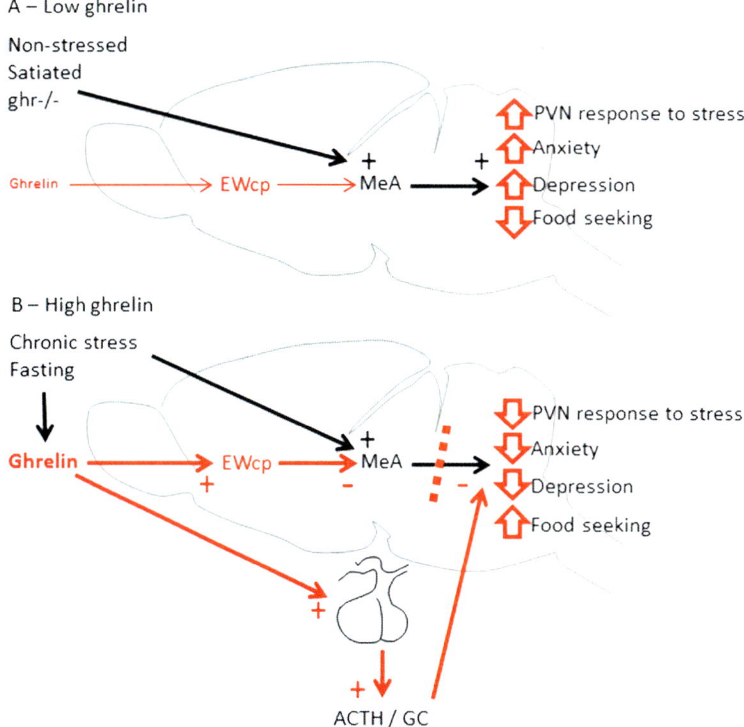

Fig. 1 Ghrelin's proposed dual role in stress, anxiety, and depression. When circulating ghrelin is low (such as in ghr-/- animals, under non-stressed conditions, with low exogenous ghrelin, and in satiety) animals are likely to display greater responses to acute stress, higher levels of anxiety and depression and reduced food-seeking (A). When ghrelin is high (such as after chronic stress, fasting, or with high exogenous ghrelin) stress, anxiety and depressive behaviours are attenuated and food-seeking promoted (B). We hypothesize this is due to ghrelin's excitatory influence on the growth hormone secretagogue receptor (GHSR) at the centrally projecting Edinger Westphall nucleus (EWcp) leading to an inhibitory effect on the medial amygdala (MeA) that would otherwise promote anxiety, depression and paraventricular nucleus of the hypothalamus (PVN) responses to stress. Ghrelin also facilitates adrenocorticotropic hormone (ACTH) release from the anterior pituitary and therefore glucocorticoid (GC) negative feedback to attenuate stress responses under conditions of stress. Ghrelin pathways indicated in *red*

learning and memory (Finger et al. 2012). This work begs the question whether obese subjects will also be resistant to ghrelin's effects on stress and mood. If this is the case, it may partly explain the significantly higher incidence of stress-related disorders in obesity. From a health perspective, if the obese are resistant to ghrelin's effects on stress, anxiety and depression, correcting the obesity with weight loss or otherwise restoring the sensitivity of these pathways to ghrelin could restore appropriate behavioural and HPA axis responses to stress. An alternative scenario is that there is some specificity to ghrelin resistance due to ghrelin's actions at extra-hypothalamic regions being maintained during obesity. In particular, we have seen at least part of ghrelin's role in regulating stress lies outside the brain, at the pituitary (Spencer et al. 2012). If we can establish ghrelin resistance in obesity is specific to food intake, ghrelin could be immensely useful in ameliorating anxiety and depression in the obese without increasing eating.

One of the chief concerns with using ghrelin receptor ligands as therapeutic agents in non-obese subjects is that they stimulate feeding (Andrews 2010). However, the desacylated form of ghrelin does not act at the ghrelin receptor and does not stimulate food intake (Neary et al. 2006), some studies suggest that it may even suppress hunger (Asakawa et al. 2005) and may therefore be an excellent candidate as a therapeutic agent against stress-related mood disorders. These possibilities remain exciting avenues for future study in this area.

Acknowledgments This work was supported by a Discovery Project Grant from the Australian Research Council (ARC; DP130100508), and Project Grant from the National Health and Medical Research Council (APP1011274). SJS is an ARC Future Fellow (FT110100084) and an RMIT University VC Senior Research Fellow.

References

Abiles V, Rodriguez-Ruiz S, Abiles J, Mellado C, Garcia A, Perez De La Cruz A, Fernandez-Santaella MC (2010) Psychological characteristics of morbidly obese candidates for bariatric surgery. Obes Surg 20:161–167

Alvarez-Crespo M, Skibicka KP, Farkas I, Molnar CS, Egecioglu E, Hrabovszky E, Liposits Z, Dickson SL (2012) The amygdala as a neurobiological target for ghrelin in rats: neuroanatomical, electrophysiological and behavioral evidence. PLoS ONE 7:e46321

Andrews ZB (2010) The extra-hypothalamic actions of ghrelin on neuronal function. Trends Neurosci 34:31–40

Anxiety and Depression Association of America (2013) http://www.adaa.org/about-adaa/press-room/facts-statistics

Arvat E, Maccario M, Di Vito L, Broglio F, Benso A, Gottero C, Papotti M, Muccioli G, Dieguez C, Casanueva FF, Deghenghi R, Camanni F, Ghigo E (2001) Endocrine activities of ghrelin, a natural growth hormone secretagogue (GHS), in humans: comparison and interactions with hexarelin, a nonnatural peptidyl GHS, and GH-releasing hormone. J Clin Endocrinol Metab 86:1169–1174

Asakawa A, Inui A, Kaga T, Yuzuriha H, Nagata T, Fujimiya M, Katsuura G, Makino S, Fujino MA, Kasuga M (2001) A role of ghrelin in neuroendocrine and behavioral responses to stress in mice. Neuroendocrinology 74:143–147

Asakawa A, Inui A, Fujimiya M, Sakamaki R, Shinfuku N, Ueta Y, Meguid MM, Kasuga M (2005) Stomach regulates energy balance via acylated ghrelin and desacyl ghrelin. Gut 54:18–24

Australian Health Survey (2011) Australian bureau of statistics. National survey of mental health and wellbeing: summary of results, 2007

Barim AO, Aydin S, Colak R, Dag E, Deniz O, Sahin I (2009) Ghrelin, paraoxonase and arylesterase levels in depressive patients before and after citalopram treatment. Clin Biochem 42:1076–1081

Berton O, Mcclung CA, Dileone RJ, Krishnan V, Renthal W, Russo SJ, Graham D, Tsankova NM, Bolanos CA, Rios M, Monteggia LM, Self DW, Nestler EJ (2006) Essential role of BDNF in the mesolimbic dopamine pathway in social defeat stress. Science 311:864–868

Block JP, He Y, Zaslavsky AM, Ding L, Ayanian JZ (2009) Psychosocial stress and change in weight among US adults. Am J Epidemiol 170:181–192

Briggs DI, Enriori PJ, Lemus MB, Cowley MA, Andrews ZB (2010) Diet-induced obesity causes ghrelin resistance in arcuate NPY/AgRP neurons. Endocrinology 151:4745–4755

Cabral A, Suescun O, Zigman JM, Perello M (2012) Ghrelin indirectly activates hypophysiotropic CRF neurons in rodents. PLoS ONE 7:e31462

Carlini VP, Monzon ME, Varas MM, Cragnolini AB, Schioth HB, Scimonelli TN, De Barioglio SR (2002) Ghrelin increases anxiety-like behavior and memory retention in rats. Biochem Biophys Res Commun 299:739–743

Carlini VP, Varas MM, Cragnolini AB, Schioth HB, Scimonelli TN, De Barioglio SR (2004) Differential role of the hippocampus, amygdala, and dorsal raphe nucleus in regulating feeding, memory, and anxiety-like behavioral responses to ghrelin. Biochem Biophys Res Commun 313:635–641

Chuang JC, Zigman JM (2010) Ghrelin's roles in stress, mood, and anxiety regulation. Int J Pept 2010. doi:10.1155/2010/460549

Cryan JF, Slattery DA (2007) Animal models of mood disorders: recent developments. Curr Opin Psychiatry 20:1–7

Currie PJ (2003) Integration of hypothalamic feeding and metabolic signals: focus on neuropeptide Y. Appetite 41:335–337

Dallman MF (2009) Stress-induced obesity and the emotional nervous system. Trends Endocrinol Metab 21:159–165

Doyle AC, Le Grange D, Goldschmidt A, Wilfley DE (2007) Psychosocial and physical impairment in overweight adolescents at high risk for eating disorders. Obesity (Silver Spring) 15:145–154

Elias CF, Saper CB, Maratos-Flier E, Tritos NA, Lee C, Kelly J, Tatro JB, Hoffman GE, Ollmann MM, Barsh GS, Sakurai T, Yanagisawa M, Elmquist JK (1998) Chemically defined projections linking the mediobasal hypothalamus and the lateral hypothalamic area. J Comp Neurol 402:442–459

Emul HM, Serteser M, Kurt E, Ozbulut O, Guler O, Gecici O (2007) Ghrelin and leptin levels in patients with obsessive-compulsive disorder. Prog Neuropsychopharmacol Biol Psychiatry 31:1270–1274

Finger BC, Dinan TG, Cryan JF (2012) Diet-induced obesity blunts the behavioural effects of ghrelin: studies in a mouse-progressive ratio task. Psychopharmacology 220:173–181

Florentzson M, Svensson K, Astin-Nielsen M, Andersson K, Hakanson R, Lindstrom E (2009) Low gastric acid and high plasma gastrin in high-anxiety Wistar Kyoto rats. Scand J Gastroenterol 44:401–407

Furness JB, Hunne B, Matsuda N, Yin L, Russo D, Kato I, Fujimiya M, Patterson M, Mcleod J, Andrews ZB, Bron R (2011) Investigation of the presence of ghrelin in the central nervous system of the rat and mouse. Neuroscience 193:1–9

Gibson EL (2006) Emotional influences on food choice: sensory, physiological and psychological pathways. Physiol Behav 89:53–61

Goldstone AP, Prechtl De Hernandez CG, Beaver JD, Muhammed K, Croese C, Bell G, Durighel G, Hughes E, Waldman AD, Frost G, Bell JD (2009) Fasting biases brain reward systems towards high-calorie foods. Eur J Neurosci 30:1625–1635

Hansson C, Haage D, Taube M, Egecioglu E, Salome N, Dickson SL (2011) Central administration of ghrelin alters emotional responses in rats: behavioural, electrophysiological and molecular evidence. Neuroscience 180:201–211

Hansson C, Annerbrink K, Nilsson S, Bah J, Olsson M, Allgulander C, Andersch S, Sjodin I, Eriksson E, Dickson SL (2013) A possible association between panic disorder and a polymorphism in the preproghrelingene. Psychiatry Res 206:22–25

Heinrichs SC, Menzaghi F, Pich EM, Hauger RL, Koob GF (1993) Corticotropin-releasing factor in the paraventricular nucleus modulates feeding induced by neuropeptide Y. Brain Res 611:18–24

Hosoda H, Kangawa K (2012) Standard sample collections for blood ghrelin measurements. Methods Enzymol 514:113–126

Ishitobi Y, Kohno K, Kanehisa M, Inoue A, Imanaga J, Maruyama Y, Ninomiya T, Higuma H, Okamoto S, Tanaka Y, Tsuru J, Hanada H, Isogawa K, Akiyoshi J (2012) Serum ghrelin levels and the effects of antidepressants in major depressive disorder and panic disorder. Neuropsychobiology 66:185–192

Johnstone LE, Srisawat R, Kumarnsit E, Leng G (2005) Hypothalamic expression of NPY mRNA, vasopressin mRNA and CRF mRNA in response to food restriction and central administration of the orexigenic peptide GHRP-6. Stress 8:59–67

Juruena MF, Cleare AJ (2007) Overlap between atypical depression, seasonal affective disorder and chronic fatigue syndrome. Rev Bras Psiquiatr 29(Suppl 1):S19–S26

Kanehisa M, Akiyoshi J, Kitaichi T, Matsushita H, Tanaka E, Kodama K, Hanada H, Isogawa K (2006) Administration of antisense DNA for ghrelin causes an antidepressant and anxiolytic response in rats. Prog Neuropsychopharmacol Biol Psychiatry 30:1403–1407

Kern A, Albarran-Zeckler R, Walsh HE, Smith RG (2012) Apo-ghrelin receptor forms heteromers with DRD2 in hypothalamic neurons and is essential for anorexigenic effects of DRD2 agonism. Neuron 73:317–332

Kloiber S, Ising M, Reppermund S, Horstmann S, Dose T, Majer M, Zihl J, Pfister H, Unschuld PG, Holsboer F, Lucae S (2007) Overweight and obesity affect treatment response in major depression. Biol Psychiatry 62:321–326

Kluge M, Schussler P, Schmid D, Uhr M, Kleyer S, Yassouridis A, Steiger A (2009) Ghrelin plasma levels are not altered in major depression. Neuropsychobiology 59:199–204

Kluge M, Schussler P, Dresler M, Schmidt D, Yassouridis A, Uhr M, Steiger A (2011) Effects of ghrelin on psychopathology, sleep and secretion of cortisol and growth hormone in patients with major depression. J Psychiatr Res 45:421–426

Kojima M, Kangawa K (2006) Drug insight: The functions of ghrelin and its potential as a multitherapeutic hormone. Nat Clin Pract Endocrinol Metab 2:80–88

Kristensson E, Sundqvist M, Hakanson R, Lindstrom E (2007) High gastrin cell activity and low ghrelin cell activity in high-anxiety Wistar Kyoto rats. J Endocrinol 193:245–250

Kristenssson E, Sundqvist M, Astin M, Kjerling M, Mattsson H, Dornonville De La Cour C, Hakanson R, Lindstrom E (2006) Acute psychological stress raises plasma ghrelin in the rat. Regul Pept 134:114–117

Kurt E, Guler O, Serteser M, Cansel N, Ozbulut O, Altinbas K, Alatas G, Savas H, Gecici O (2007) The effects of electroconvulsive therapy on ghrelin, leptin and cholesterol levels in patients with mood disorders. Neurosci Lett 426:49–53

Lloyd RB, Nemeroff CB (2011) The role of corticotropin-releasing hormone in the pathophysiology of depression: therapeutic implications. Curr Top Med Chem 11:609–617

Lutter M, Sakata I, Osborne-Lawrence S, Rovinsky SA, Anderson JG, Jung S, Birnbaum S, Yanagisawa M, Elmquist JK, Nestler EJ, Zigman JM (2008) The orexigenic hormone ghrelin defends against depressive symptoms of chronic stress. Nat Neurosci 11:752–753

Malik S, Mcglone F, Bedrossian D, Dagher A (2008) Ghrelin modulates brain activity in areas that control appetitive behavior. Cell Metab 7:400–409

Mitchell V, Bouret S, Beauvillain JC, Schilling A, Perret M, Kordon C, Epelbaum J (2001) Comparative distribution of mRNA encoding the growth hormone secretagogue-receptor (GHS-R) in Microcebus murinus (Primate, lemurian) and rat forebrain and pituitary. J Comp Neurol 429:469–489

Nakashima K, Akiyoshi J, Hatano K, Hanada H, Tanaka Y, Tsuru J, Matsushita H, Kodama K, Isogawa K (2008) Ghrelin gene polymorphism is associated with depression, but not panic disorder. Psychiatr Genet 18:257

Neary NM, Druce MR, Small CJ, Bloom SR (2006) Acylated ghrelin stimulates food intake in the fed and fasted states but desacylated ghrelin has no effect. Gut 55:135

Nestler EJ, Hyman SE (2010) Animal models of neuropsychiatric disorders. Nat Neurosci 13:1161–1169

Novick JS, Stewart JW, Wisniewski SR, Cook IA, Manev R, Nierenberg AA, Rosenbaum JF, Shores-Wilson K, Balasubramani GK, Biggs MM, Zisook S, Rush AJ (2005) Clinical and demographic features of atypical depression in outpatients with major depressive disorder: preliminary findings from STAR*D. J Clin Psychiatry 66:1002–1011

Ochi M, Tominaga K, Tanaka F, Tanigawa T, Shiba M, Watanabe T, Fujiwara Y, Oshitani N, Higuchi K, Arakawa T (2008) Effect of chronic stress on gastric emptying and plasma ghrelin levels in rats. Life Sci 82:862–868

Oliver G, Wardle J (1999) Perceived effects of stress on food choice. Physiol Behav 66:511–515

Otto B, Tschop M, Heldwein W, Pfeiffer AF, Diederich S (2004) Endogenous and exogenous glucocorticoids decrease plasma ghrelin in humans. Eur J Endocrinol 151:113–117

Pani L, Porcella A, Gessa GL (2000) The role of stress in the pathophysiology of the dopaminergic system. Mol Psychiatry 5:14–21

Papadimitriou A, Priftis KN (2009) Regulation of the hypothalamic-pituitary-adrenal axis. NeuroImmunoModulation 16:265–271

Patterson ZR, Ducharme R, Anisman H, Abizaid A (2010) Altered metabolic and neurochemical responses to chronic unpredictable stressors in ghrelin receptor-deficient mice. Eur J Neurosci 32:632–639

Patterson ZR, Khazall R, Mackay H, Anisman H, Abizaid A (2013) Central ghrelin signaling mediates the metabolic response of C57BL/6 male mice to chronic social defeat stress. Endocrinology 154:1080–1091

Petersen PS, Woldbye DP, Madsen AN, Egerod KL, Jin C, Lang M, Rasmussen M, Beck-Sickinger AG, Holst B (2009) In vivo characterization of high Basal signaling from the ghrelin receptor. Endocrinology 150:4920–4930

Pinar M, Gulsun M, Tasci I, Erdil A, Bolu E, Acikel C, Doruk A (2008) Maprotiline induced weight gain in depressive disorder: changes in circulating ghrelin and adiponectin levels and insulin sensitivity. Prog Neuropsychopharmacol Biol Psychiatry 32:135–139

Raspopow K, Abizaid A, Matheson K, Anisman H (2010) Psychosocial stressor effects on cortisol and ghrelin in emotional and non-emotional eaters: influence of anger and shame. Horm Behav 58:677–684

Richardson LP, Davis R, Poulton R, Mccauley E, Moffitt TE, Caspi A, Connell F (2003) A longitudinal evaluation of adolescent depression and adult obesity. Arch Pediatr Adolesc Med 157:739–745

Rouach V, Bloch M, Rosenberg N, Gilad S, Limor R, Stern N, Greenman Y (2007) The acute ghrelin response to a psychological stress challenge does not predict the post-stress urge to eat. Psychoneuroendocrinology 32:693–702

Sacher J, Neumann J, Funfstuck T, Soliman A, Villringer A, Schroeter ML (2012) Mapping the depressed brain: a meta-analysis of structural and functional alterations in major depressive disorder. J Affect Disord 140:142–148

Sapolsky RM, Romero LM, Munck AU (2000) How do glucocorticoids influence stress responses? Integrating permissive, suppressive, stimulatory, and preparative actions. Endocr Rev 21:55–89

Schanze A, Reulbach U, Scheuchenzuber M, Groschl M, Kornhuber J, Kraus T (2008) Ghrelin and eating disturbances in psychiatric disorders. Neuropsychobiology 57:126–130

Schellekens H, Finger BC, Dinan TG, Cryan JF (2012) Ghrelin signalling and obesity: at the interface of stress, mood and food reward. Pharmacol Ther 135:316–326

Schellekens H, Dinan TG, Cryan JF (2013) Ghrelin at the interface of obesity and reward. Vitam Horm 91:285–323

Schmid DA, Wichniak A, Uhr M, Ising M, Brunner H, Held K, Weikel JC, Sonntag A, Steiger A (2006) Changes of sleep architecture, spectral composition of sleep EEG, the nocturnal secretion of cortisol, ACTH, GH, prolactin, melatonin, ghrelin, and leptin, and the DEX-CRH test in depressed patients during treatment with mirtazapine. Neuropsychopharmacology 31:832–844

Schneiderman N, Ironson G, Siegel SD (2005) Stress and health: psychological, behavioral, and biological determinants. Annu Rev Clin Psychol 1:607–628

Scott KM, Mcgee MA, Wells JE, Oakley Browne MA (2008) Obesity and mental disorders in the adult general population. J Psychosom Res 64:97–105

Serlachius A, Hamer M, Wardle J (2007) Stress and weight change in university students in the United Kingdom. Physiol Behav 92:548–553

Simon GE, Von Korff M, Saunders K, Miglioretti DL, Crane PK, Van Belle G, Kessler RC (2006) Association between obesity and psychiatric disorders in the US adult population. Arch Gen Psychiatry 63:824–830

Spencer SJ, Tilbrook A (2011) The glucocorticoid contribution to obesity. Stress 14:233–246

Spencer SJ, Xu L, Clarke MA, Lemus M, Reichenbach A, Geenen B, Kozicz T, Andrews ZB (2012) Ghrelin regulates the hypothalamic-pituitary-adrenal axis and restricts anxiety after acute stress. Biol Psychiatry 72:457–465

Staufenbiel SM, Penninx BW, Spijker AT, Elzinga BM, Van Rossum EF (2012) Hair cortisol, stress exposure, and mental health in humans: a systematic review. Psychoneuroendocrinology 38:1220–1235

Strasser F (2012) Clinical application of ghrelin. Curr Pharm Des 18:4800–4812

Takaya K, Ariyasu H, Kanamoto N, Iwakura H, Yoshimoto A, Harada M, Mori K, Komatsu Y, Usui T, Shimatsu A, Ogawa Y, Hosoda K, Akamizu T, Kojima M, Kangawa K, Nakao K (2000) Ghrelin strongly stimulates growth hormone release in humans. J Clin Endocrinol Metab 85:4908–4911

Torres SJ, Nowson CA (2007) Relationship between stress, eating behavior, and obesity. Nutrition 23:887–894

Vieweg WV, Julius DA, Fernandez A, Tassone DM, Narla SN, Pandurangi AK (2006) Posttraumatic stress disorder in male military veterans with comorbid overweight and obesity: psychotropic, antihypertensive, and metabolic medications. Prim Care Companion J Clin Psychiatry 8:25–31

Yamanaka A, Beuckmann CT, Willie JT, Hara J, Tsujino N, Mieda M, Tominaga M, Yagami K, Sugiyama F, Goto K, Yanagisawa M, Sakurai T (2003) Hypothalamic orexin neurons regulate arousal according to energy balance in mice. Neuron 38:701–713

Zheng J, Dobner A, Babygirija R, Ludwig K, Takahashi T (2009) Effects of repeated restraint stress on gastric motility in rats. Am J Physiol Regul Integr Comp Physiol 296:R1358–R1365

Zigman JM, Elmquist JK (2003) Minireview: from anorexia to obesity–the yin and yang of body weight control. Endocrinology 144:3749–3756

Zigman JM, Jones JE, Lee CE, Saper CB, Elmquist JK (2006) Expression of ghrelin receptor mRNA in the rat and the mouse brain. J Comp Neurol 494:528–548

Ghrelin and Parkinson's Disease

Marcus M. Unger and Wolfgang H. Oertel

Abstract The peptide ghrelin regulates gastrointestinal motility and energy homeostasis. Ghrelin is also a modulator of higher brain functions like mood, cognition, sleep, and reward-associated behaviour. Some of these functions regulated by ghrelin are disturbed in the neurodegenerative disorder Parkinson's disease. The link between ghrelin and Parkinson's disease is further endorsed by the finding that ghrelin receptors are expressed in brain regions that undergo neurodegeneration in Parkinson's disease and by the finding of an altered postprandial ghrelin secretion in patients with Parkinson's disease. In addition, ghrelin has shown protective effects in neurodegenerative disorders including experimental models of Parkinson's disease. This chapter reviews the potential link between the gastric peptide ghrelin and the movement disorder Parkinson's disease.

Keywords Parkinson's disease · Neurodegeneration · Neuroprotection · Dopaminergic neurons · Deep brain stimulation · Postprandial ghrelin secretion, growth hormone secretagogue receptor 1a (ghrelin receptor 1a)

Parkinson's Disease

Parkinson's disease (PD) is a common neurodegenerative disorder of the elderly. The disease manifests by slowness of movements (bradykinesia), muscle rigidity, tremor at rest, and gait disturbance. Besides motor impairments, PD is accompanied by a number of non-motor symptoms including neuropsychiatric disturbances

M. M. Unger (✉)
Department of Neurology, Saarland University, Kirrberger Strasse,
66421 Homburg, Germany
e-mail: marcus.unger@uks.eu

W. H. Oertel
Department of Neurology, Philipps-University Marburg, Marburg, Germany
e-mail: oertelw@med.uni-marburg.de

J. Portelli and I. Smolders (eds.), *Central Functions of the Ghrelin Receptor*,
The Receptors 25, DOI: 10.1007/978-1-4939-0823-3_13,
© Springer Science+Business Media New York 2014

and gastrointestinal symptoms. Two pathohistological hallmarks are the presence of intracellular protein aggregates (Lewy bodies) and the degeneration of dopaminergic neurons that project from the brainstem (substantia nigra pars compacta) to the basal ganglia (dorsal striatum). The disruption of this pathway results in typical parkinsonian motor symptoms. Yet, Parkinson-related neurodegeneration is neither restricted to nigrostriatal neurons nor to the dopaminergic system. Besides the nigro-striatal pathway there is also neuronal loss in other (mono-aminergic and cholinergic) brain regions and degenerative changes also occur outside the central nervous system (CNS), e.g. in the enteric nervous system.

Relevance of Ghrelin and Ghrelin Receptors for Parkinson's Disease

The relevance of ghrelin and ghrelin receptors for the neurological disorder PD is based on the following considerations:

Ghrelin receptors are expressed in a number of CNS regions that undergo neurodegeneration in PD (substantia nigra, dorsal nucleus of the vagal nerve, etc.) (Guan et al. 1997; Zigman et al. 2006).

Ghrelin modulates gastrointestinal motility and higher brain functions (mood, cognition, sleep, and reward-associated behaviour) (Diano et al. 2006; Dickson et al. 2011; Kluge et al. 2009, 2010; Lutter et al. 2008; Steiger et al. 2011). Gastrointestinal motility and the above-mentioned neuropsychological functions are frequently altered in patients with PD.

Another link between the neuropeptide ghrelin and PD is the vagal nerve. The vagal nerve is dysfunctional already in early stages of PD due to neurodegeneration in the corresponding nucleus in the brainstem (Braak et al. 2003). On the other hand, the gastric secretion of ghrelin is co-regulated by the vagal nerve (Masuda et al. 2000; Williams et al. 2003). In consequence, disruption of the neuronal brain–gut-axis in PD is likely to affect the gastric secretion of ghrelin.

From a therapeutic point of view, ghrelin and ghrelin receptor agonists are candidates for disease-modification in PD (due to ghrelin's neuroprotective potential (Andrews et al. 2009)) and for symptomatic treatment of certain non-motor features of PD.

Postprandial Ghrelin Secretion in Parkinson's Disease

Based on the observation that ghrelin modulates biological functions that are frequently disturbed in PD (see above), we investigated postprandial ghrelin secretion after a standardised test meal in patients at different stages of PD and in healthy volunteers (Unger et al. 2011). Healthy volunteers showed relatively high fasting ghrelin serum concentrations that dropped after the test meal and consecutively recovered within the next few hours. In patients with PD, this dynamic pattern of physiological ghrelin secretion was less pronounced. Patients with PD had descriptively lower fasting ghrelin concentrations and showed a significantly reduced recovery of ghrelin concentrations in the late postprandial phase. Interestingly, this pattern of altered ghrelin secretion was also seen in patients with idiopathic rapid-eye-movement (REM) sleep behaviour disorder, a population considered at risk or even at a pre-motor stage of PD. The inter-individual variability of ghrelin concentrations in our study was high (in controls as well as in patients with PD). In order to reassess the data of our pilot study, we performed a second study (unpublished data) in an independent cohort. In this study we also distinguished between the two subforms of ghrelin: acyl and des-acyl ghrelin. We reproduced the descriptive differences between PD patients and controls of our pilot study (especially the concentrations of acyl ghrelin were descriptively lower), but we did not identify any statistically significant differences between patients and controls. The underlying pathophysiology of the assumed disturbed ghrelin secretion in PD remains speculative. An intact vagal nerve is crucial for physiological ghrelin release. Disruption of the brain–gut-axis in PD is therefore one explanation for the observed alterations in ghrelin release. Changes in the enteric nervous system in PD might also affect production and release of ghrelin. Concerning downstream effects of reduced ghrelin secretion in PD, the finding of reduced growth hormone (GH) concentrations in patients with PD (compared to age-matched controls) (Bellomo et al. 1991) might be related to the reduced ghrelin concentrations we observed in our study (as ghrelin induces GH release).

Taken together, there is preliminary evidence that ghrelin secretion is disturbed in PD. Taking into account ghrelin's neuroprotective potential reduced ghrelin concentrations might render dopaminergic neurons more vulnerable in subjects predisposed to develop PD.

Deep Brain Stimulation (DBS) in Parkinson's Disease and Its Effect on Ghrelin

Deep brain stimulation (DBS) of the subthalamic nucleus (STN) is an established and effective therapy for patients in advanced stages of PD. DBS modulates the neuronal activity of the STN which results in improved motor abilities. The STN is located close to ghrelin-producing neurons of the hypothalamus. STN-DBS might therefore affect local secretion of ghrelin. Based on ghrelin's known orexigenic properties and the clinical observation that PD patients frequently gain weight after STN-DBS, Corcuff et al. investigated the effect of STN-DBS on circulating ghrelin concentrations (Corcuff et al. 2006). The authors did not find a significant effect of STN-DBS on circulating ghrelin concentrations in accordance with observations of other groups (Arai et al. 2012; Novakova et al. 2011). Yet, this does not exclude local effects of STN-DBS on hypothalamic neurons and a consecutive increase in locally produced ghrelin. Indeed, ghrelin is mainly produced by the stomach and the analysis of circulating ghrelin concentrations might not reflect changes in the local ghrelin production of hypothalamic neurons. In contrast to the above-mentioned studies, one recently published study reported an increase in circulating ghrelin concentrations (associated with weight gain) up to 6 months after STN-DBS (Markaki et al. 2012). The divergent observations concerning the effects of STN-DBS on ghrelin concentrations can be explained by the low number of cases investigated, the high inter-individual variability of ghrelin concentrations and differences in the analytical determination of ghrelin.

To further elucidate the effect of STN-DBS on ghrelin, cerebrospinal fluid (CSF) ghrelin concentration might be a more relevant parameter than circulating ghrelin concentrations. CSF ghrelin concentrations might better reflect changes in ghrelin produced locally in the CNS. We have recently shown the feasibility of measuring ghrelin in human CSF samples in a reliable and reproducible way (Unger et al. 2013). Due to ghrelin's neuroprotective potential, the effect of STN-DBS is not only of scientific interest but might also have clinical relevance and therapeutical implications.

Neuroprotective Effects of Ghrelin on Dopaminergic Neurons

Based on previously reported neuroprotective properties of ghrelin and based on the fact that ghrelin receptor 1a is expressed on substantia nigra dopaminergic neurons (Jiang et al. 2008; Zigman et al. 2005), several groups have investigated the effect of ghrelin on dopaminergic neurons after exposure to neurotoxins, i.e. in experimental models of PD.

Experimental dopaminergic neurodegeneration can be induced by the neurotoxin 1-methyl-4-phenyl-1,2,3,6-tetrahydropyridine (MPTP). MPTP is metabolized into 1-methyl-4-phenylpyridinium (MPP+) by the enzyme monoamine oxidase B (MAO-B) in the brain. MPP+ interferes with oxidative phosphorylation in the mitochondria of dopaminergic neurons. By this means, MPP+ disturbs the generation of new adenosine triphosphate (ATP) which leads to energy depletion and finally to neuronal death.

Moon et al. showed that peripherally administered ghrelin protects dopaminergic neurons in the MPTP mouse model of PD in a dose-dependent manner (Moon et al. 2009). Ghrelin's neuroprotective effects were mediated by suppression of matrix metalloproteinase-3 release from dopaminergic neurons and the consecutive inhibition of microglia activation. Ghrelin-treated animals showed reduced dopaminergic cell loss and preserved striatal dopamine levels after MPTP-exposure. In addition, ghrelin partially preserved motor function in MPTP-intoxicated animals. A ghrelin receptor 1a antagonist (D-Lys-3-GHRP-6) did not affect dopaminergic neurons when administered alone but reversed the neuroprotective effects of ghrelin. The ghrelin receptor 1a is, therefore, likely to be the primary mechanism by which ghrelin mediates its neuroprotective effects.

Anti-apoptotic effects are another mechanism by which ghrelin acts neuroprotective: Jiang et al. showed that ghrelin increases anti-apoptotic proteins and consecutively attenuates caspase-3 activity induced by MPTP in dopaminergic neurons of the substantia nigra pars compacta (Jiang et al. 2008). Similar to the study by Moon and colleagues, the protective effects of ghrelin (mediated via anti-apoptotic mechanisms) could be abolished by a ghrelin receptor 1a antagonist (D-Lys-3-GHRP-6). These experiments show again that neuroprotection by ghrelin is mediated via ghrelin receptors 1a.

Andrews and colleagues focused on another aspect of ghrelin's neuroprotective properties (Andrews et al. 2009). The authors investigated mitochondrial mechanisms leading to neuroprotection under conditions of cellular stress. First, Andrews et al. showed that ghrelin binds to dopaminergic neurons in the substantia nigra pars compacta and increases the neuronal activity (firing rate) of these neurons which in turn results in increased dopamine concentrations in the striatum. The authors also showed that ghrelin protects dopaminergic neurons and renders them resistant to cellular stress (neurotoxins, MPTP) by enhancing mitochondrial uncoupling protein 2 (UCP2) and thereby lowering reactive oxygen species. In contrast to the studies by Moon et al. and Jiang et al. (who used a ghrelin receptor 1a antagonist to show that the protective effects of ghrelin are mediated by the ghrelin receptor 1a), Andrews and colleagues used ghrelin receptor knockout mice to prove that ghrelin's neuroprotective properties are mediated via this specific receptor.

The three above-mentioned studies show that ghrelin exerts different effects that eventually result in protection of dopaminergic neurons under conditions of cellular stress. All these mechanisms are likely to be mainly mediated by the ghrelin receptor.

Summary

In summary, a number of clinical and experimental data endorse the relevance of the gastric peptide ghrelin for the movement disorder PD:

- Ghrelin receptors are expressed in brain regions that are prone to neurodegeneration in PD.
- Ghrelin is a modulator of gastrointestinal motility and neuropsychological functions (mood, cognition, sleep). Several of these functions are frequently altered in PD.
- Postprandial ghrelin secretion is altered in PD.
- Ghrelin exerts neuroprotective effects in experimental models of PD by binding to the ghrelin receptor 1a.

Whether or not the above-mentioned findings will eventually result in new therapeutic options for PD remains to be seen.

References

Andrews ZB, Erion D, Beiler R, Liu ZW, Abizaid A, Zigman J, Elsworth JD, Savitt JM, DiMarchi R, Tschoep M et al (2009) Ghrelin promotes and protects nigrostriatal dopamine function via a UCP2-dependent mitochondrial mechanism. J Neurosci 29:14057–14065

Arai E, Arai M, Uchiyama T, Higuchi Y, Aoyagi K, Yamanaka Y, Yamamoto T, Nagano O, Shiina A, Maruoka D et al (2012) Subthalamic deep brain stimulation can improve gastric emptying in Parkinson's disease. Brain 135:1478–1485

Bellomo G, Santambrogio L, Fiacconi M, Scarponi AM, Ciuffetti G (1991) Plasma profiles of adrenocorticotropic hormone, cortisol, growth hormone and prolactin in patients with untreated Parkinson's disease. J Neurol 238:19–22

Braak H, Del Tredici K, Rub U, de Vos RA, Jansen Steur EN, Braak E (2003) Staging of brain pathology related to sporadic Parkinson's disease. Neurobiol Aging 24:197–211

Corcuff JB, Krim E, Tison F, Foubert-Sanier A, Guehl D, Burbaud P, Cuny E, Baillet L, Gin H, Rigalleau V et al (2006) Subthalamic nucleus stimulation in patients with Parkinson's disease does not increase serum ghrelin levels. Br J Nutr 95:1028–1029

Diano S, Farr SA, Benoit SC, McNay EC, da Silva I, Horvath B, Gaskin FS, Nonaka N, Jaeger LB, Banks WA et al (2006) Ghrelin controls hippocampal spine synapse density and memory performance. Nat Neurosci 9:381–388

Dickson SL, Egecioglu E, Landgren S, Skibicka KP, Engel JA, Jerlhag E (2011) The role of the central ghrelin system in reward from food and chemical drugs. Mol Cell Endocrinol 340:80–87

Guan XM, Yu H, Palyha OC, McKee KK, Feighner SD, Sirinathsinghji DJ, Smith RG, Van der Ploeg LH, Howard AD (1997) Distribution of mRNA encoding the growth hormone secretagogue receptor in brain and peripheral tissues. Brain Res Mol Brain Res 48:23–29

Jiang H, Li LJ, Wang J, Xie JX (2008) Ghrelin antagonizes MPTP-induced neurotoxicity to the dopaminergic neurons in mouse substantia nigra. Exp Neurol 212:532–537

Kluge M, Gazea M, Schussler P, Genzel L, Dresler M, Kleyer S, Uhr M, Yassouridis A, Steiger A (2010) Ghrelin increases slow wave sleep and stage 2 sleep and decreases stage 1 sleep and REM sleep in elderly men but does not affect sleep in elderly women. Psychoneuroendocrinology 35:297–304

Kluge M, Schussler P, Schmid D, Uhr M, Kleyer S, Yassouridis A, Steiger A (2009) Ghrelin plasma levels are not altered in major depression. Neuropsychobiology 59:199–204

Lutter M, Sakata I, Osborne-Lawrence S, Rovinsky SA, Anderson JG, Jung S, Birnbaum S, Yanagisawa M, Elmquist JK, Nestler EJ et al (2008) The orexigenic hormone ghrelin defends against depressive symptoms of chronic stress. Nat Neurosci 11:752–753

Markaki E, Ellul J, Kefalopoulou Z, Trachani E, Theodoropoulou A, Kyriazopoulou V, Constantoyannis C (2012) The role of ghrelin, neuropeptide Y and leptin peptides in weight gain after deep brain stimulation for Parkinson's disease. Stereotact Funct Neurosurg 90:104–112

Masuda Y, Tanaka T, Inomata N, Ohnuma N, Tanaka S, Itoh Z, Hosoda H, Kojima M, Kangawa K (2000) Ghrelin stimulates gastric acid secretion and motility in rats. Biochem Biophys Res Commun 276:905–908

Moon M, Kim HG, Hwang L, Seo JH, Kim S, Hwang S, Lee D, Chung H, Oh MS, Lee KT et al (2009) Neuroprotective effect of ghrelin in the 1-methyl-4-phenyl-1,2,3,6-tetrahydropyridine mouse model of Parkinson's disease by blocking microglial activation. Neurotox Res 15:332–347

Novakova L, Haluzik M, Jech R, Urgosik D, Ruzicka F, Ruzicka E (2011) Hormonal regulators of food intake and weight gain in Parkinson's disease after subthalamic nucleus stimulation. Neuro Endocrinol Lett 32:437–441

Steiger A, Dresler M, Schussler P, Kluge M (2011) Ghrelin in mental health, sleep, memory. Mol Cell Endocrinol 340:88–96

Unger MM, Moller JC, Mankel K, Eggert KM, Bohne K, Bodden M, Stiasny-Kolster K, Kann PH, Mayer G, Tebbe JJ et al (2011) Postprandial ghrelin response is reduced in patients with Parkinson's disease and idiopathic REM sleep behaviour disorder: a peripheral biomarker for early Parkinson's disease? J Neurol 258:982–990

Unger MM, Oertel WH, Tackenberg B (2013) Cerebrospinal fluid concentrations of ghrelin in patients with multiple sclerosis. Neuro Endocrinol Lett 34:14–17

Williams DL, Grill HJ, Cummings DE, Kaplan JM (2003) Vagotomy dissociates short- and long-term controls of circulating ghrelin. Endocrinology 144:5184–5187

Zigman JM, Jones JE, Lee CE, Saper CB, Elmquist JK (2006) Expression of ghrelin receptor mRNA in the rat and the mouse brain. J Comp Neurol 494:528–548

Zigman JM, Nakano Y, Coppari R, Balthasar N, Marcus JN, Lee CE, Jones JE, Deysher AE, Waxman AR, White RD et al (2005) Mice lacking ghrelin receptors resist the development of diet-induced obesity. J Clin Invest 115:3564–3572

Index

J. Portelli and I. Smolders (eds.), *Central Functions of the Ghrelin Receptor*,
The Receptors 25, DOI: 10.1007/978-1-4939-0823-3,
© Springer Science+Business Media New York 2014